Otto Seifert, Friedrich Müller

Manual of clinical diagnosis

Otto Seifert, Friedrich Müller

Manual of clinical diagnosis

ISBN/EAN: 9783743348318

Manufactured in Europe, USA, Canada, Australia, Japa

Cover: Foto ©berggeist007 / pixelio.de

Manufactured and distributed by brebook publishing software (www.brebook.com)

Otto Seifert, Friedrich Müller

Manual of clinical diagnosis

MANUAL

OF

CLINICAL DIAGNOSIS

BY

Dr. OTTO SEIFERT AND Dr. FRIEDRICH MÜLLER

PRIVATDOCENT IN WÜRZBURG   ASSISTENT DER II. MED. KLINIK IN BERLIN

THIRD EDITION

REVISED AND CORRECTED BY

Dr. FRIEDRICH MÜLLER

TRANSLATED, WITH THE PERMISSION OF THE AUTHORS, BY

WILLIAM BUCKINGHAM CANFIELD, A.M., M.D. (BERLIN)

Fellow of the American Academy of Medicine; Member of the Medical and Chirurgical
Faculty of Maryland ; Visiting Physician to the Union Protestant Infirmary
of Baltimore; Lecturer on Normal Histology, and Chief of Clinic for
Throat and Chest, University of Maryland.

WITH SIXTY ILLUSTRATIONS

NEW YORK & LONDON

G. P. PUTNAM'S SONS

𝔗𝔥𝔢 𝔎𝔫𝔦𝔠𝔨𝔢𝔯𝔟𝔬𝔠𝔨𝔢𝔯 𝔓𝔯𝔢𝔰𝔰

1887

PREFACE TO THE FIRST EDITION.

THE presentation of this manual to the public is due to the encouragement of our highly esteemed teacher and *chef*, Geheimrath Professor Gerhardt. We have endeavored to supply a want by giving in an epitomized form, the different methods of examination, as well as a convenient collection of those data and figures which should always be familiar to the physician and student. These data, on account of their number and variety, cannot be remembered with the necessary exactness, and, on the other hand, are so scattered throughout numberless text-books and monographs, that it would be troublesome and time-wasting to search for them. In selecting and arranging this material, we have been led by the experience gained in holding courses, and we have also endeavored to consider the practical needs of the student and physician by noting only what is reliable, and omitting every thing self-evident and of secondary importance.

THE AUTHORS.

WÜRZBURG and BERLIN, April, 1886.

PREFACE TO THE THIRD EDITION.

In preparing the third edition of this manual, I have endeavored to do justice to all the wishes expressed by the different critics, as well as to consider any wants which have become apparent since the last edition. Consequently, a number of improvements and additions have been made, and among them it seemed necessary to add some new illustrations, especially to the chapters on blood and urine. The illustrations of the leucocytes are from preparations of Professor Ehrlich, and those of the urinary sediment are, in part, taken from the physico-chemical atlas of Funke. The tables in the last chapter are intended to make the questions of diet and assimilation of practical use in the sick-room. In conclusion, I should like to thank all those gentlemen who have so kindly assisted us by their suggestions.

BERLIN, October, 1886.

TRANSLATOR'S PREFACE.

THE favor with which this book has been received in Germany, and its eminently practical and concise manner of dealing with the different important points in diagnosis, seem to justify its translation into English. It has been brought down to the latest acquisitions of science, thus representing the most advanced views. For the sake of clearness, the figures relating to weight, measure, length, etc., as well as the dose table at the end of the book, have been modified to conform to the system used in America and England. Translations from the original into French and Russian are now in press.

The translator takes great pleasure in thanking in this place his friend Dr. Robert T. Wilson, for kind services and valuable suggestions rendered in the proof-reading and correction.

W. B. C.

1010 NORTH CHARLES ST., BALTIMORE,
September, 1887.

CONTENTS.

CLINICAL DIAGNOSIS.

CHAPTER I.

THE BLOOD.

THE whole *quantity of blood* in the body of an adult is equal to about $\frac{1}{13}$ of the weight of the body—that is, on an average, 5 kilograms [10 lbs.].[1]

The *specific gravity* varies in health between 1045 and 1075.

The *reaction* of the blood is alkaline.

The *amount of hæmoglobin*[2] in the blood is about 14.57 grams [4 drachms] in men, and 13.27 grams [3¼ drachms] in women, in 100 ccm. [3 ounces] of blood. On heating, the hæmoglobin is resolved into brown hæmatin and albumen.

If some blood (*e. g.*, that obtained from a blood stain on wood or linen), be heated to the boiling point with glacial acetic acid and a trace of common salt, and then slowly evaporated, there are formed brownish-yellow rhombic crystals of the muriate of *hæmatin*, which is the same thing as *hæmin* or Teichmann's crystals. The preparation should then be moistened with a little glycerine, and examined with a high power under the microscope.

The *red blood corpuscles* measure in healthy individuals

---

[1] That part enclosed in [ ] is by the translator.

[2] The amount of hæmoglobin is determined by the quantitative spectral analysis or by means of a hæmochromomcter.

from 6.7 μ to 9.3 μ.[1]   The average size is 7.8 μ (Gram).
Giant blood corpuscles are found principally in the blood
of the anæmic, and especially in those suffering from
progressive pernicious anæmia.   The *dwarf blood cor
puscles* measure from 2.2 to 6 μ, and are like the normal
ones, only slightly more biconcave.   These are also
found frequently in anæmia.

By *Poikilocytes* are meant those red corpuscles of
irregular form (pear-, club-, or biscuit-shaped) which
are seen in all anæmic conditions.   *Microcytes* are small
spherical bodies, generally very rich in hæmoglobin, and

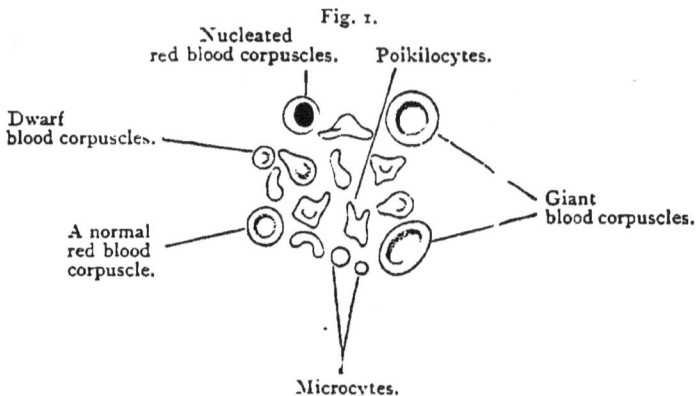

Fig. 1.

are found in many cases of burning and poisoning.   Still
we must very often consider them as artificial products.
It is uncertain whether red corpuscles with crenated
edges (thorn-apple-shaped corpuscles) appear in normal
blood or not.   When seen, they are generally considered
artificial products caused by evaporation.   Notwith-
standing this, they are observed to form more quickly
and abundantly in many cachectic conditions than in
normal blood.

---

[1] μ = the one-thousandth part of a millimetre, and is known as a
micro-millimetre or micron, and equals about $\frac{1}{25600}$ of an inch.

Nucleated red blood corpuscles are seen in all severe ·cases of anæmia but they can only be recognized in stained preparations.[1] Very large nucleated blood corpuscles (megalocytes) are seen in progressive per- nicious anæmia.

Blood-plaques (Bizzozero[2]) = *Hæmatoblasts* (Hayem[3]) are colorless flat round discs about one half the diameter of a red blood corpuscle. They change their shapes very quickly when outside of the blood-vessel. The *elementary granular masses* are small, often angular, colorless granules with a diameter of $1-2\mu$. They consist in part of fat, and are probably for the most part disin- tegrated products of the blood-plaques.

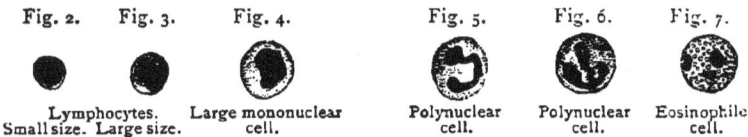

Fig. 2.	Fig. 3.	Fig. 4.	Fig. 5.	Fig. 6.	Fig. 7.
Lymphocytes. Small size.	Large size.	Large mononuclear cell.	Polynuclear cell.	Polynuclear cell.	Eosinophile cell.

The *white blood corpuscles* (Leucocytes) are divided according to Ehrlich into : (1) *Lymphocytes*, which are about the size of a red blood corpuscle or somewhat larger, with a large round nucleus, and a very small, often scarcely visible, protoplasm. We distinguish two forms, a smaller and a larger, which latter is only the former in a more progressive stage of development. The lymphocytes have their origin in the lymphatic glands. (2) *Large mononuclear forms*, with large round or oval nucleus and broad protoplasmic body. These are the earlier stages of development of the third. (3) *The large polynuclear form*, containing a nucleus very much divided and lobulated, and which may be deeply stained

---

[1] *Fortschritte der Medicin*, 1884.

[2] Bizzozero : *Virchow's Archiv.*, Bd. xc.

[3] Hayem : *Archive de Physiologie*, 1878–79.

with aniline colors. They form by far the greatest number of the leucocytes, and are found almost exclusively in pus. Ehrlich designates as *eosinophile cells* those leucocytes in whose protoplasm a quantity of coarse, fatty, shining granules are seen, which are colored an intense red on staining a preparation of dried blood with a one-per-cent. watery solution of eosin. These cells have their origin in the marrow of bones, and are present in normal blood in small quantities only. In myelogenetic leucæmia they are seen in large numbers. In lymphatic leucæmia it is principally the small lymphocytes which are increased in number

In order to examine the blood, it is generally sufficient to cleanse and dry the finger, and, with a needle or lancet, to make a quick and deep puncture in the tip. The drop of blood should then escape without squeezing the finger, and be dropped on a clean cover-glass, which is, in turn, dropped on the slide in such a way that the blood is spread out in a thin film. In order to see the blood-plaques, a drop of a one-per-cent. osmic-acid solution is applied to the finger, and the puncture is made through this drop. Instead of the osmic acid, which is simply a conservative fluid, a thin watery solution of methyl violet with 0.6 % of common salt may be used, which colors the blood-plaques and the nuclei of the nucleated red blood-corpuscles. In finer examinations of the blood, it is better to color a dried preparation (*vid*. Chap. xi. for preparation) with the following solution :

R   Hæmatoxylin, 2 grams [30 grains].
    Alcohol,
    Glycerine,
    Distilled water, āā 100 grams [3 ounces].
    Glacial acetic acid, 10 grams [2½ drachms].
    Alum in excess.

This mixture should stand 3 weeks in the light, and a few granules of eosin should be added to it. The dry preparations remain from 6–12 hours in the staining fluid, and then are to be washed off with water and examined. The nuclei of the nucleated red blood-corpuscles will be found to be stained intensely black. (Ehrlich.)

The *number of red blood corpuscles* is, on an average, in men, in the normal condition, 5 million ; in women, 4½ million, to a cubic millimetre [a millimetre equals $\frac{1}{25}$ of an inch].

˙ The *number of white blood corpuscles* varies between 5,000–10,000, and is temporarily increased after a hearty meal. The number of blood-plaques is about 200,000 to the cubic millimetre.

The proportion between white and red blood corpuscles is, in healthy individuals, 1–500 to 1–1,000. A proportion which is more than 1 to 400 must be considered as a pathological increase in the number of white corpuscles.

Welker and Moleschott considered the proportion of white to red corpuscles to be 1:330 and 1.357. Dupérié found 5.500,000, red to 5,000 white, or 1:1100 ; Malassez, 1:1200 ; Hayem, Bouchut, Dubrisay found, on an average, 1:500–1,000 ; Halla, 1:422–811. Laache and Otto found, on an average, for men 4.97 and 4.99 million, and for women 4.43 and 4.58 million red blood corpuscles.

In order to count the corpuscles, a deep puncture is made into the finger-tip, and the escaping drop is sucked up into the mélangeur until it reaches the mark 1. The point of the instrument is then wiped off, and the *diluting fluid* is sucked up to the mark 101. This mixture of blood and diluting fluid [1] is well shaken and introduced into the counting chamber, and covered by the cover-glass, which should be lightly pressed on, and then the corpuscles are counted in each square, which is etched on the cover-glass. If a thousand corpuscles have been counted in one square, the amount of corpuscles in a cubic millimetre can be calculated, since the dilution of the blood (1:100) and the depth of the chamber are known. By using the chamber of Thoma-Zeiss (depth $\frac{1}{10}$ mm., 1 square $= \frac{1}{100}$ cubic millimetre), the average number of corpuscles in a small square

---

[1] This fluid may be either a 3 % solution of common salt, or a 5 % solution of Glauber's salt, or Hayem's solution, which is corrosive sublimate, 0.5 grams [7 grains], Glauber's salt, 5.0 [1¼ drachms], common salt, 2.0 [½ drachm], distilled water 200.0 grams [6 ounces].

is multiplied by 400,000—that is, the whole number of corpuscles is to be divided by the number of squares which have been counted. It is more convenient to count the four small squares in one column and calculate the average result of a large number of counts. This number, which is the number of blood corpuscles in any four squares, is then multiplied by 100,000. In using the chamber of Malassez or Hayem, which has a depth of $\frac{1}{5}$ mm., the average number which is in a large rectangle (= 20 small squares) is to be multiplied by 10,000. If the blood dilution is 1:200 instead of 1:100, *i. e.* (up to mark 0.5 of the mélangeur), the result should be multiplied by 2.

Leucocytosis, that is, an increase of the leucocytes in proportion to the red corpuscle, is observed in numerous acute diseases (typhus abdominalis, erysipelas, etc.), also in cachectic conditions (cancer). This increase of leucocytes is often very great, and may even reach as high as 1 to 60 red corpuscles.

In *Leucæmia*, the amount of red corpuscles as well as of hæmoglobin is generally considerably decreased, so that the number of leucocytes is very considerably increased, and almost equals that of the red corpuscles ; indeed it may equal or exceed it. In the first stages of the disease, when the increase of the white corpuscles is often less than in severe *leucocytosis*, the diagnosis of leucæmia can be certain only when, in its further course, a rapid increase of leucocytes makes it very evident ; or when the proportion of white to red exceeds 1:50. In myelogenetic leucæmia, there are numerous eosinophile leucocytes, and also nucleated red blood corpuscles to be seen. *Lymphatic* leucæmia is characterized by an increase of *lymphocytes*.

In *pseudoleucæmia* there is a slight decrease in the number of red corpuscles, and in the amount of hæmoglobin, and no increase of the leucocytes.

In the first few days after heavy loss of blood, the amount of red corpuscles as well as of the hæmoglobin sinks markedly to over 50 % of the normal, whereas the number of leucocytes increases. In the period of convalescence, the amount of red corpuscles increases more quickly than that of the hæmoglobin. In the *secondary anæmiæ*, after typhus abdominalis, tuberculosis, malaria, lead poisoning, ankylostomiasis, nephritis, cancer, etc., the number of red corpuscles as well as the amount of hæmoglobin is diminished, and the amount of white blood corpuscles increased (leucocytosis).

In *chlorosis* the amount of hæmoglobin is very greatly decreased, whereas the number of red corpuscles is often very little or not at all increased. These are therefore very pale. The amount of white corpuscles is normal.

In *progressive pernicious anæmia*, the number of red corpuscles is enormously reduced, often to $\frac{1}{10}$ of the normal, whereas their size, and above all things, the amount of hæmoglobin is increased.

An *increase* in the number of red corpuscles is observed in thickening of the blood in cholera, as well as in many heart affections.

After long-existing malaria, *pigment-containing leucocytes* are at times seen in the blood.

Micro-organisms are also observed in the blood—*e. g., tubercle bacilli* in miliary tuberculosis, *bacilli lepræ, bacilli anthracis*, and the *spirilla of recurrent fever.* The latter can be seen with medium power, and are best recognized from the fact that when they come in contact with the red blood corpuscles, they impart to them a jerking motion ; or they can be recognized by coloring them as a dried preparation, with a watery solution of gentian violet, as in the case of the bacilli anthracis.

CHAPTER II.

TEMPERATURE.

THE temperature of the body is generally taken either in the axillary space or in the rectum [and under the tongue]. In the rectum it is about 0.5°–1° higher than in the axilla.

The temperature of the *healthy individual* measures [1] in the axilla between 36.2° C. [97.1° F.] and 37.5° C. [99.5° F.]. The highest temperature is late in the afternoon, and the lowest, very early in the morning. An elevation of temperature can temporarily occur in consequence of bodily exertion, taking food, hot-baths, etc. A continuous elevation of temperature occurs *in fever*. According to Wunderlich we have :

The temperature of collapse, 36° C. [96.8° F.].

Sub-febrile temperature 37.5°–38° C. [99.5° F.–100.4° F.].

Slight fever 38°–38.5° C. [100.4°–101.3° F.].

Moderate fever 39° C. [102.2° F.] morning ; 39.5° [103.1° F.] evening.

Considerable fever 39.5° C. [103.1° F.] morning ; 40.5° C. [104.9° F.] evening.

High fever over 39.5°C. [103.1° F.] morning ; over 40.5° C. [104.9° F.] evening.

---

[1] In order to convert from one scale to the other, the following formula may be used :
$$N° C = \tfrac{4}{5} n° \; R = \tfrac{9}{5} n° + 32° \; F.$$

Hyperpyrexia, or fever over 41.5° C. [106.7°].

Also in fever there is usually a morning remission and an evening exacerbation. Exceptionally, especially in phthisis, we have the reverse—typus inversus. The difference between the highest and lowest temperature decides its type of the fever, thus:

Febris continua = a daily difference of not more than 1° C. [1.8° F.].

Febris remittens = a daily difference of not more than 1.5° C. [2.7° F.].

Febris intermittens = in the course of the day the high temperature is varied by a period of no fever.

In the course of a fever we distinguish:

I. *Stadium incrementi* = a quick rise of temperature, generally accompanied by a chill or a slowly rising temperature.

II. *Fastigium* = or a stage of highest temperature. Its transition to the next stage is known as the amphibolic stage.

III. *Stadium decrementi.* The fever fall can follow either slowly, in course of several days, in which case we have lysis; or quickly, the crisis. At the actual crisis the temperature falls rapidly (in one day) until it goes below normal. This fall is generally accompanied by a profuse perspiration. A high rise of temperature often precedes the crisis, which is called perturbatio critica.

In acute infectious diseases we distinguish the *stage of incubation*—that is, the time between the moment of contagion and the outbreak of the disease. Also in acute exanthematous diseases there is the *prodromal stage*, or stage of the first morbid appearances, *i. e.*, before the outbreak of the eruption.

Morbilli—Measles.

Incubation, ten days. Prodromal stage, three days, characterized by affections of the mucous membranes.

Fig. 8.
Temperature chart in Morbilli.

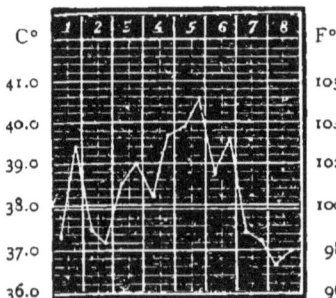

Prodromes. Eruption. Defervescence.

It begins with chill and high fever ; and on the second or third day there is a slight fall of temperature. At the appearance of the eruption (on the face) the temperature rises again, and reaches its highest point when the eruption is most widely disseminated. This is the stadium floritionis, and lasts three to four days. The critical fall of temperature occurs on the sixth or seventh day of the disease, after which desquamation begins.

Scarlatina—Scarlet-Fever.

Incubation, four to seven days. Prodromal stage, one to two days. It is characterized by angina. It begins with a chill and quick rise of temperature. At the end of the first or second day there is an outbreak of the eruption (on the breast), and as it spreads the temperature rises. Defervescence begins on fourth to seventh day of the disease, and comes to an end slowly, with the paling of the eruption in three to six days later. Desquamation then follows.

Fig. 9.
Temperature chart in Scarlatina.

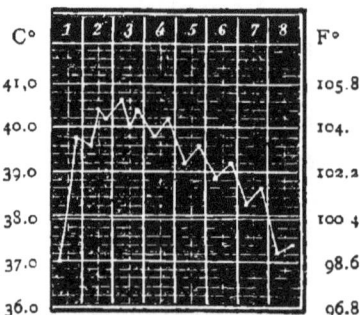

Variola—Small-pox.

Incubation, nine days (nine to sixteen days), at the end of which time there is general disturbance of functions. The prodromal stage (two to five days) begins with a chill, with sudden rise of temperature, and often on the second or third day the first signs of the eruption are observed. With the be-

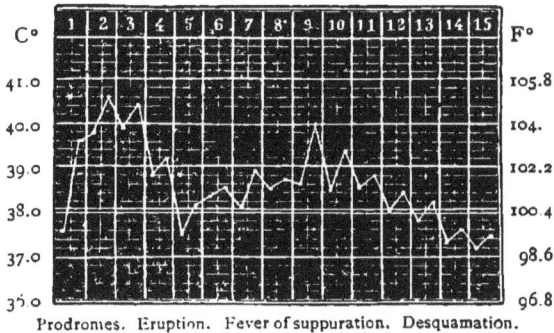

Fig. 10. Temperature chart in Variola.

Prodromes. Eruption. Fever of suppuration. Desquamation.

ginning of the pustulation there is a quick fall of temperature. Then comes a second and in the beginning a slight febrile movement, which reaches its climax on about the ninth day (fever of suppuration), preserves a remittent type for some time, and after a varying length ends in lysis (period of desquamation). About the sixteenth day the stadium decrustationis begins.

Variolois—Varioloid.

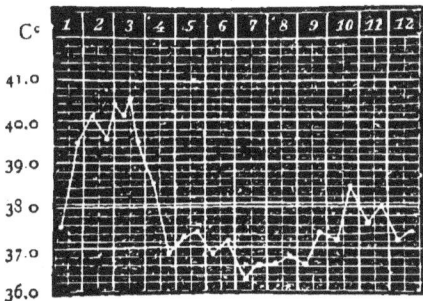

Fig. 11. Temperature chart in Variolois.

The incubation and prodromal stages are the same as in variola vera, only much lighter. The second period of fever (fever of suppuration) is wanting. The period of desquamation often begins as early as the ninth or tenth day, and is generally accompanied by slight rise of temperature.

Varicella—Chicken-pox.

The prodromes are generally wanting. The eruption of the vesicles begins with slight fever. The vesicles dry up after three or four days.

Typhus Abdominalis— Typhoid Fever.

Incubation, seven to twenty-one days. The prodromal stage lasts several days to a week, and is accompanied by general disturbances. In the first week of the disease the temperature rises by degrees, accompanied by slight chills, and reaches, on the fourth or seventh day, its highest point. This continues as a febris continua until the third week in the milder forms, and until the fifth week in the more severe forms. Then the morning temperature begins gradually to fall, while the evening temperature still remains high, until gradually lysis results, which in mild cases is in the fourth week. There is also

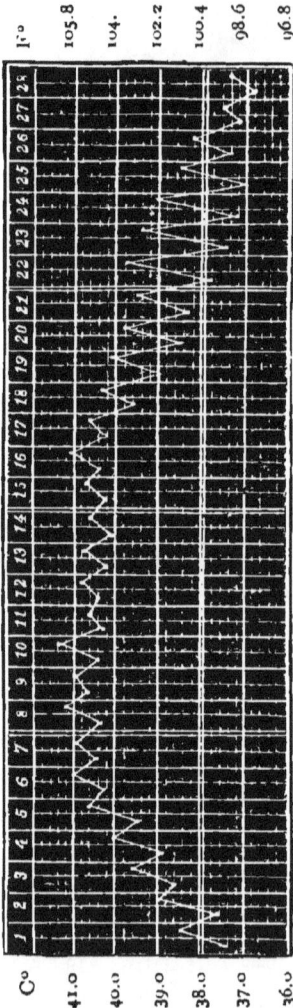

Fig. 12.   Temperature chart in Typhus abdominalis.

tumefaction of the spleen in the second half of the first week of the disease. Roseola occurs on the sixth to ninth day of the disease.

Typhus Exanthematicus—Typhus Fever.

Incubation varies from a few days to three weeks. The prodromal stage is not marked. The disease begins with a chill and rapid rise of temperature, and then it remains febris continua until the thirteenth to the seven-

Fig. 13. Temperature chart in Typhus exanthematicus.

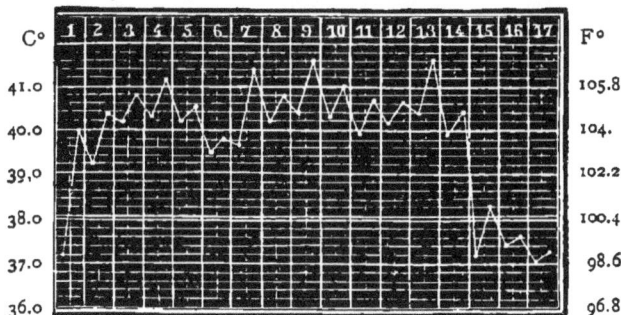

Eruption.

teenth day. There are often remissions at the end of the first week. There is then a critical fall of temperature, at times with transitory perturbatio critica. The eruption appears on the third to the sixth day after an inflammation of the mucous tracts.

Febris Recurrens—Recurrent Fever.

Incubation five to seven days. The prodromal stage is not clearly marked. The fever begins with violent chill and a (high) sudden rise of temperature, which continues as in febris continua until the fifth or seventh day, and then critically falls. After a period of apyrexia, lasting about a week, there is again an attack of fever as at first, but not lasting so long. Often, after a period of seven days, there is a third attack, lasting one to two days.

Fig. 14.   Temperature chart in Febris recurrens.

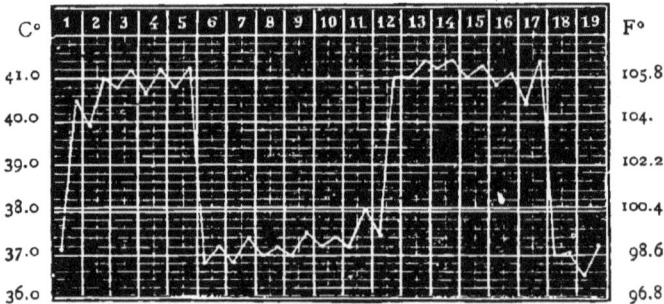

Malaria—Febris Intermittens.

Incubation seven to twenty-one days. Prodromal stage is not marked. There is a chill, and the temperature rises to very great height, and then sinks after a few hours to or below the normal. There is also strong perspiration. According as the fever occurs every day, or every second and third day, it is called quotidian, tertian, and quartan intermittent fever. Febris intermittens duplicata is that type in which two attacks occur in quick succession in the course of the same day. Febris intermittens anteponens and post-ponens is that type in which the new attack of fever does not generally occur at the same hour of the day as the preceding attack, but sooner or later.

Fig. 15.

Temperature chart in Febris intermittens.

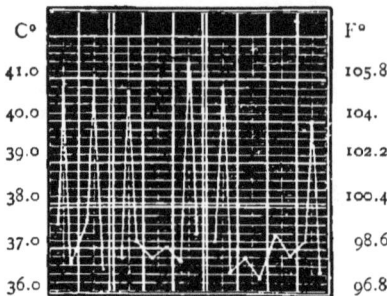

Febris quotid;   tertiana;   quartana.

Erysipelas.

Incubation one to eight days. It generally begins
with a chill and high temper-
ature. On the first or second
day an inflammation of the
skin is observed. The temperature continues high as
long as the morbid process
spreads, and quickly falls as
soon as the erysipelas ceases
spreading. With the spas-
modic spread of the erup-
tion there is often found a remittent or intermittent fever.

Fig. 16.
Temperature chart in
Erysipelas.

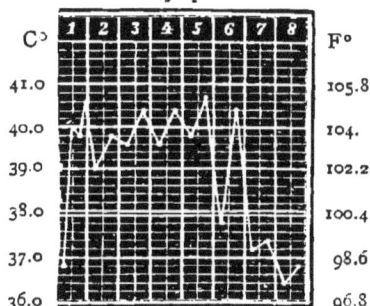

Pneumonia Crouposa.

It begins with a chill and sudden rise of temperature.
There is a continued fever during the spread of the
peneumonic infiltration. On the fifth to the seventh
day, and at times later, the crisis occurs, generally very

Fig. 17.   Fig. 18.
Temperature charts in Pneumonia crouposa.

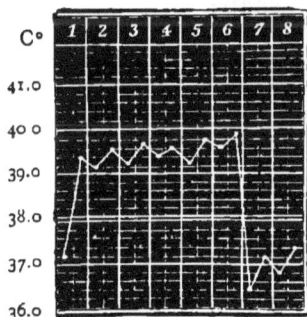

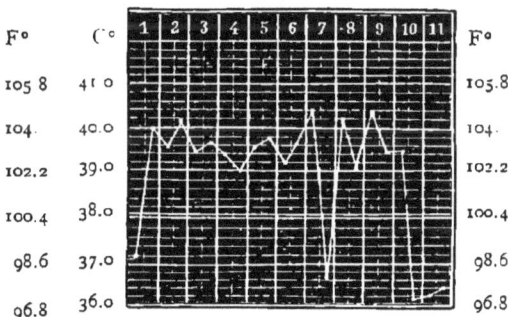

suddenly, with strong perspiration, and, at the same time,
a decrease in the frequency of the pulse and respiration.
There is often a pseudo-crisis (v. Fig. 18) one or two
days before the real crisis, but here the pulse and respiration remain high.

CHAPTER III.

ORGANS OF RESPIRATION

Topography of the Chest.

The *vertebral column* has a normal curvature at the cervical, dorsal, lumbar, and sacral regions. A pathological curvature of the vertebral column convexly backward is called *cyphosis*. The same deformity, not curved but angular in character, is called *gibbus*. A curvature forward is called *lordosis*, and a lateral curvature *scoliosis*. A deforming curvature both laterally and backward is called *cypho-scoliosis*.

The *sternum* in the adult is about 16–20 cm. [6–8 inches] long. The angular prominence between the manubrium and body is called the angulus Ludovici. A bending inward of the xiphoid process, and of the lower part of the body, is called funnel breast. This latter shape of the thorax is *acquired*, and is seen in some occupations, *e. g.*, in cobblers who press their instruments against the breast (cobbler's breast). When the costal cartilages in rachitis are pressed in by lateral compression, and the sternum has a keel shape, it is called pectus carinatum or chicken breast.

The *clavicle* has a supra- and infra-clavicular groove in it. The external part of the latter is called Mohrenheim's groove.

The *scapula* covers the second to seventh or third to eighth rib, and is provided with a fossa supraspinata and a fossa infraspinata. Between the inner border of each scapula is the inter-scapular space.

In order to determine the *height of the thorax* in front, we must follow the ribs by beginning to count at the second; and behind the landmarks are the spinous processes, beginning with the seventh cervical, the vertebra prominens.

The *Harrison furrow* has a horizontal direction at the level of the xiphoid process, and corresponding to the normal attachment of the diaphragm. The region below this furrow to the angle of the ribs is called the hypochondrium.

In order to determine the *breadth of the thorax*, we make use of the following perpendicular lines :

(1) The median line.

(2) The parasternal line, drawn half-way between the border of the sternum and the nipple.

(3) The mammary line, drawn through the nipple, which, in healthy adults, lies between the fourth and fifth ribs, 10 cm [4 inches] distant from the border of the sternum.

(4) The anterior, middle, and posterior axillary line ; the former drawn through the anterior, the latter through the posterior, boundary of the axilla.

(5) The scapular line, drawn through the lower border of the scapula.

The costo-articular line is a line drawn from the sterno-costal articulation to the tip of the eleventh rib.

Size of the Thorax. The sterno-vertebral diameter of the thorax measures, in healthy men, 16.5 cm. [6½ inches] above, and 19.2 cm. [7½ inches] below, and in women it is somewhat smaller. The broad diameter of the chest is, in men, at the height of the nipple, 26 cm. [10 inches].

The circumference of the chest at the height of the nipple measures, in healthy men, 82.0 cm. [32½ inches], after deep expiration, and 90 cm. [36 inches] after deep inspiration—that is to say, the maximum excursion of the thorax in respiration is 8.0 cm [3 inches]. In those who are right-handed, the circumference of the right half of the chest is 0.5 to 2.0 cm [¼ to 1 inch] larger than the left. In left-handed persons the left side is slightly larger.

Enlargement of one or both sides of the chest is observed in pneumothorax, in effusions into the pleural sac (occasionally in pneumonia), often in mediastinal tumors, and in emphysema. In the latter disease there is, in severe cases, a barrel-shaped chest, while all the diameters, but especially the sterno-vertebral diameter, are enlarged, so that there results a permanent position of inspiration. Enlargements, especially of the lower opening of the thorax, occur with tumors and effusions in the peritoneal cavity.

A *narrow thorax* may be congenital or acquired. A *congenital* narrowing of the thorax, in which it is long, small, shallow, while the intercostal spaces are broad and the sterno-vertebral diameter is especially smaller, is called a *paralytic shape* of the thorax, and is most frequent in phthisis pulmonum.

An acquired narrowing of the thorax may be caused by an absorption of a pleuritic exudation and shrinkage of the lungs, as in phthisis and cirrhosis pulmonum.

The *number of respirations* in the healthy adult is from 16 to 20, and in new-born children 44, a minute.

The *normal relation between the frequency of respiration and pulse* is as 1:3½ to 4.

The inspiratory enlargement of the thorax takes place in the male, principally by a deep descent of the abdomen, and partly by the raising of the ribs by the scaleni and intercostal muscles — *typus costo-abdominalis.* In women, the inspiration is carried on more by raising the ribs—*typus costalis.*

The *expiratory narrowing* of the thorax is, under normal conditions, caused only by the elasticity of the chest, without muscular assistance.

Inspiration and expiration are generally of the same duration, and follow each other without the intervention of a pause.

The lungs perform no active movement during respiration, but passively follow the movements of the chest wall and the diaphragm. In healthy individuals at rest, infrequent and superficial respirations are sufficient to change the air in the lungs, but as soon as the amount of carbonic acid gas becomes too great in the lungs, the respiration becomes more frequent and deeper. This is seen in bodily exertion and fever, and also in disturbances of the circulation in consequence of heart troubles, and in all diseased conditions of the respiratory tract itself. If the blood is overloaded with carbonic acid gas to too great an extent, difficulty of breathing, *i. e.*, *dyspnœa*, sets in.

In *inspiratory dyspnœa*, in which long-drawn inspirations are carried out with great muscular exertion, while the expirations follow more easily, the accessory muscles come into play. The sternomostoidei, scaleni, levatores costarum, serratus posticus superior, serratus anticus major, pectoralis major and minor, levator scapulæ, trapezius, rhomboidei major and minor, the extensors of the vertebral columns, the dilators of the nasal and oral cavities as well as of the larynx. This kind of dyspnœa is observed in narrowing of the air passages—*e. g.*, of the larynx, the trachea, and the bronchi, as well as in many diseased conditions of the lungs where there is a decrease of the respiratory surface.

In severe inspiratory dyspnœa, there is an *inspiratory drawing in* in the region of the xiphoid process and the lower border of the ribs.

In *expiratory dyspnœa*, in which the narrowing of the chest is rendered difficult, and the length of the expiration is increased in proportion to that of the inspiration, the abdominal muscles and the serratus posticus inferior and the quadratus lumborum come into play as the accessory muscles of respiration.   Expiratory dyspnœa is observed in cases of laryngeal polypus, more especially in *emphysema* and *bronchial asthma.*

The *mixed* form of dyspnœa is made up of the inspiratory and expiratory dyspnœa.

Changes in the Frequency of Breathing.

Increase in respiratory frequency takes place :

1. From nervous causes, in all affections of the mind, and in hysteria.

2. From accumulation of carbonic acid gas in the blood, from bodily exertion, in fever, and in many forms of heart disease.

3. In most diseases of the respiratory tract, such as pneumonia, phthisis, pleurisy, emphysema, accumulation of fluid and liquid in the pleural cavity, and finally, in all diseases of the abdomen—*e. g.*, in peritonitis, tumors, ascites, which hinder the movements of the diaphragm.

The relation between the frequency of the pulse and respiration may thus be changed from 1:4 to 1:1.

Retarding of the respiration is observed in stenosis of the upper air-passages, and in cerebral diseases (as in hemorrhages, tumors, etc.).

The *Cheyne-Stokes respiration* is a kind of breathing in which periods of complete cessation from breathing (apnœa) are varied with periods of slowly rising respiratory movements, which become gradually deeper, and then, in turn, fall.   This phenomenon is observed in many severe cerebral diseases, in heart disease, and uræmia.

Spirometry.

The *total capacity of the lungs*, *i. e.*, that quantity of air which, after forced inspiration, can be expelled by forced expiration, is about 3,000–4,000 ccm. [200–250 cubic inches] in a healthy man, or an average of 3,600 ccm. [230 cubic inches] ; in woman, 2,000–3,000 ccm. [125–200 cubic inches], or an average of 2,500 ccm. [163 cubic inches].   This capacity increases with the growth of the body, so that about 22 ccm. [1⅓ cubic inches] of expired air would be equivalent to about 1 cm. [⅓ of an inch] in the adult. The capacity of the lungs is not so great in children, in old men, in all diseases of the stomach, and when the stomach is full.

The *complemental air* is that amount of air which, after quiet inspiration, can be introduced by forced inspiration, and equals 1,500 ccm. [100 cubic inches].

The *reserve air* is that amount of air which, after quiet expiration, can be expelled by forced expiration, equal to about 1,500 ccm. [100 cubic inches].

The *ordinary breathing air* is that amount of air which is introduced and expelled in quiet respiration, and is equal to about 500 ccm. [33 cubic inches].

The *residual air* is that amount of air which remains in the lungs after the deepest exspiration, and equals about 1,600–2,000 ccm. [100–125 cubic inches].

Percussion of the Thorax.

In percussion, the following qualities of sound are distinguished :

1. Clear and dull.
(2. Full and empty.)
3. Tympanitic and non-tympanitic.
4. High and deep.

Besides these, we have the metallic sound and cracked-pot sound (bruit de pot fêlé).

In the normal thorax there is a clear sound over the lungs, and a dull sound over the organs not containing air.

The Normal Boundaries of the Lung.

The *upper boundary* of the lungs (apex) is, in front, 3–4 cm. [1–1½ inch] above the upper border of the clavicle, and behind, it is on a level with the spinous process of the seventh cervical vertebra.

The lower border of the lungs is, at the right border of the sternum, on a level with the sixth rib ; in the right mammary line it is at the upper border of the seventh rib ; in the anterior axillary line at the lower border of the seventh rib ; and in the scapular line at the ninth rib ; and at the vertebral column at the spinous process of the eleventh dorsal vertebra.   On the left of, and near the sternum is the cardiac dulness.   The boundary between the left lung and the tympanitic stomach is not easily defined.

Topography of the Different Lobes of the Lung.

The border between the upper and lower lobes begins behind, on both sides, at the level of the second and third dorsal vertebra, takes its course downwards and outwards, and reaches its limit on the left side in the mammary line at the sixth rib ; on the right side it divides about 6 cm. [2½ inches] above the angle of the scapula into an upper and lower branch which embrace the middle lobe.   The upper one takes a course slightly downwards, and reaches the anterior border of the lung at the height of the fourth or fifth costal cartilage ; the lower one separating the middle lobe from the lower lobe, goes straight down to reach the lower border of the lung at the mammary line.   On percussion behind, therefore, we have, on both sides, the upper lobe only to the third rib, and from there on downwards the lower lobe ; in front, on the left side, only the upper lobe, and on the right side the upper and middle lobes.

In quiet respiration the lung borders move but little ; in a supine position, the anterior lower border is about 2 cm. [1 inch] lower than in an upright position ; in lying on the side, the lower lung-border of the opposite side descends in the axillary line 3–4 cm. [1–1½ inches].   In extreme inspiration the respiratory movement can be very considerable, and in forced inspiration, while lying on the side, the displacement may be as much as 9 cm. [3½ inches]. The respiratory movement of the lungs (by filling the complemental space) is greatest in the axillary line.

In emphysema the lower border of the lung is observed to be *permanently lower*, and in asthmatic attacks, temporarily lower than in the normal.

The *lower border* of the lung is *higher than normal* in all contractions of the lung and pleura, by pressure upward of the diaphragm, as well as by collections of air, and fluids, and of tumors in the abdominal cavity. The *upper border* of the lung is *lower than normal* in shrinking of the apex in consequence of tuberculosis.

The respiratory movements are less in emphysema and brown induration of the lungs, in beginning pleurisy, and in adhesion of both pleural surfaces.

DULNESS over the lung substance may be present :

1. When that part of the lung next to the thoracic wall contains no air. Still, such a place must be at least as large as the pleximeter, and 2 cm. [1 inch] thick, in order to be recognized. The parenchyma of the lung may be without air in pneumonia and tuberculous infiltration, in hæmorrhagic infarct, abscess, neoplasm of the lungs, and in atelectasis (by compression of the lung or by an obstruction in the bronchus leading to it).

2. When a fluid or solid substance (tumor, pleuritic effusion) is between the lung and the thoracic wall ; still, fluids in adults must amount at least to 400 ccm. [25 cubic inches] in order to be found.

· Pleuritic exudations collect in a non-adherent pleural cavity, first in the posterior inferior parts, and extend from there forward and upward. If the exudation has been formed while the patient was recumbent, the upper border of the dulness forms an inclined line from behind and above to in front and below. But if the exudation arise while the patient is walking about, then the line is almost horizontal. In exudations which are undergoing absorption, the upper border often has a curved course, which is highest at the side of the thorax (curve of Damoiseau or Ellis).

In inflammatory pleural exudation the borders of the dulness change little or not at all when the patient changes his position, since the exudation is generally encapsuled by the adhesions of the pleural surface. In hydrothorax, which is generally bilateral, although not at the same height on both sides, the level of the fluid changes only after some time. In a collection of air and fluid at the same time in the pleural sac (pyo- and sero-pneumothorax) the border of the fluid becomes horizontal at once, since in the upright position of the patient the fluid can be diagnosticated as a dulness in the anterior inferior half of the thorax, while in the supine position it sinks backward, and makes room in front for the tympanitic sound.

The other organs are often displaced by the collection of large quantities of air or fluid in the pleural sac The displacement of the heart is not so great in left-sided pleural exudations as in right-sided ones.

A *tympanitic sound* in the thorax near a healthy lung is observed in the lowest part only of the left lung bordering on the stomach.

Pathologically, a *tympanitic sound* is observed :

1. In *condensation of the lung tissue*, which permits of a percussion of the column of air in the bronchi and trachea, that is, of those air conductors which are normally present in the lungs, *e. g.*, in infiltration of the upper lobe.

2. In the presence of *pathological* air-conducting cavities.

(*a*) In *cavities* which have firm walls, or in those which, with smooth walls, are separated from the thoracic cavity by infiltrated tissue, *e. g.*, bronchiectatic or tuberculous cavities, when these are at least as large as a walnut.

(*b*) In *pneumothorax*, so long as the air is not under too strong pressure ; for if the latter is the case, the sound, which was before tympanitic, will become dull and non-resonant.

3. In *relaxation of the lung tissue* in the neighborhood of extended infiltrations and of pleuritic and pericarditic exudations ; thus, for example, there is often heard a tympanitic sound over the upper lobe of a lung, when there is a pneumonia of the lower lobe, or when the lung is compressed by a pleuritic exudation.

4. In *incomplete* infiltrations of the lung tissue, when it contains air and fluid, as, for example, in the first and third stages of croupous pneumonia, in catarrhal pneumonia, and œdema of the lungs.

A *metallic sound* depends upon the prominence of the high upper tones together with the fundamental tone, and gradually decreasing reverberation. A metallic sound arises in the thorax from the presence of large smooth-walled cavities whose diameter is at least 6 cm. [2½ inches].

The *cracked-pot sound* arises on strong percussion when the air is forced through a narrow opening out of a cavity (murmur of stenosis). It is also found in healthy persons, especially in children, if the chest is percussed during speaking or crying. Pathologically it is heard over all superficial cavities which are connected with the bronchi by a narrow opening, and at times also when the parenchyma of the lung is relaxed and infiltrated. This sound is clearer on strong short percussion when the patient opens the mouth. If this sound is also tinkling, it is called metallic tinkling.

The *height and depth of the percussion note* is distinguished principally by tympanitic and metallic percussion note, and the note is deeper according as the cavity is larger, and the opening is narrow.

Wintrich's change of note is that in which the percussion note is higher when the mouth is *open*, and lower when it is closed. It is observed in cavities and pneumothorax, if they are in open communication with a bronchus, except, at times, in pneumonia and pleuritic exudation above. If this change of sound is observed on lying down, and is absent on sitting up, or *vice versa*, the bronchus leading to it is obstructed in certain positions by the fluid contents (interrupted change of tone of Wintrich).

Respiratory change of tone is at times observed over cavities, by there being a higher tone in deep inspiration.

The change of tone of Gerhardt, that is, different heights of the percussion note on *sitting* and *lying*, is observed over cavities which have unequal diameters and are partly filled with fluid. According as the patient sits upright or lies horizontally, the fluid changes its position, and thus the longest diameter of the (oval) cavity, which diameter determines the height of the tone, is made longer or shorter. We may suppose that the longest diameter is horizontal when the percussion note is deepest. The most reliable sign of the formation of a cavity is to be regarded as the deeper tone on sitting up, by which the longest diameter is directed from before behind.

The *change of note of Biermer* is the percussion note in a pneumothorax (containing also fluid) which is deeper on sitting up than on lying down, since in the former position, the diaphragm is forced down by the pressure of the fluid, and thus the sounding cavity is made larger.

Auscultation.

The Breathing Sound.

We distinguish :

1. *Vesicular.*
2. *Bronchial.*
3. *Amphoric.*
4. *Undetermined.*
5. *Metamorphosing.*

Vesicular breathing. Over the healthy lung is heard, during inspiration, a soft sucking murmur, and during expiration, a short uncertain or vesicular murmur, or none at all. In children we find an especially loud and sharp vesicular respiration : *puerile respiration.* Vesicular breathing represents a bronchial breathing which originates in the trachea and the large bronchi, and is modified by the overlying lungs. Vesicular breathing may be imitated by saying *f* or *v* softly.

A *diminution of vesicular breathing* is heard in obstruction and narrowing of the bronchi, compression of the lungs, emphysema, and also when the lungs are pressed back from the thoracic wall by fluid (pleuritis), or solid substances (tumors).

Increased vesicular breathing is observed in the swelling and narrowing of the bronchi, *e. g.*, in bronchial catarrh.

A *lengthening* and *sharpening of the vesicular expiratory tone* is heard when the exit of air from the bronchi is prevented by swelling of the mucous lining, or by an accumulation of the secretion in bronchitis and bronchial asthma. In the beginning of phthisis pulmonum, the same thing is frequently observed, but then it is confined to the apices of the lungs.

When the inspiration is interrupted by two or more pauses, we call it *jerking respiration.*

A *systolic vesicular respiration* is one in which the inspiratory murmur is strengthened with each heart beat.

Bronchial breathing, or sonorous breathing, which corresponds to the tympanitic percussion note, is heard, in healthy individuals, over the larynx and trachea and interscapular space. We can imitate this sound by pronouncing *k* [the German *ch*]. It is stronger and lasts longer in expiration than in inspiration.

Under pathological conditions, we observe this kind of breathing when the respiratory murmur which arises in the larger bronchi, or in smooth-walled cavities, is transmitted unchanged through consolidated lung tissue to the chest wall, as, for example, in pneumonia or tuberculous *infiltration*, and in *compression* of the lung above a pleural exudation, except in phthisical or bronchiectatic cavities which lie near the thoracic wall, or which are surrounded by consolidated tissue. In this case, the bronchus which leads to it must be unobstructed.

Amphoric breathing is a deep, hollow, buzzing sound, heard over cavities which give forth a metallic percussion note, *e. g.*, in large, smooth-walled cavities at least as large as the closed hand, and in pneumothorax. This sound may be imitated by blowing over a jug or a bottle.

An *undetermined respiratory murmur* is one which has neither the character of bronchial nor vesicular breathing.

Metamorphosing respiration is characterized by an inspiration which begins with vesicular breathing, and then passes over into bronchial breathing. It is heard principally over cavities.

Râles

are respiratory murmurs caused by the collection of fluid or mucus in the air passages, or by the inspiratory current of air which forces apart the adhering bronchial walls. We distinguish râles according as they are :

1. More or less plentiful.

2. Moist or dry.
3. Mucous, crepitant, or sub-crepitant.
4. Metallic or non-metallic.

Dry râles may be purring or whistling (sibilant and sonorous). These are heard in accumulation of a thick secretion, and in œdema of the mucous membrane.

Moist râles are divided into mucous, crepitant, and sub-crepitant, of which the former occur only over large cavities, the latter over smaller ones. *Subcrepitant râles* are heard on inspiration, and occur in the first and third stages of pneumonia, in œdema of the lung, as well as in persons sick or convalescent, who have been for some time in a recumbent position, in whom the crepitant râle is heard in the posterior, inferior part of the lung, during the first deep inspiration.

Metallic râles are heard under the same circumstances as bronchial breathing over consolidated lung tissue, cavities, etc.

Metallic tinkling râles of high musical pitch are heard over large cavities which give forth a metallic percussion note and amphoric breathing. To this class belongs the sound of drops of fluid falling into the cavity, as in pneumothorax.

Auscultation of the Voice.

On auscultating the chest of a healthy person while talking, the only thing heard is an indistinct murmur.

This auscultatory sign is *weakened* by *obstruction* or compression of the bronchi, or by a layer of air or fluid between the chest wall and the lung (pleurisy, pneumothorax, etc.).

Bronchophony, or increased transmission of the voice, is heard where the sound waves of the broncho-tracheal column of air are conducted through the consolidated lung tissue to the chest wall, *c. g.*, in pneumonia, cavities, above a pleuritic exudation, etc. Very strong bronchophony is called *pectoriloquy.*

A special kind of bronchophony is called *ægophony*, by which we understand a high trembling or bleating variety of voice. This is observed more frequently in incomplete compression of the bronchi at the upper border of a medium-sized pleuritic exudation, less frequently in hydrothorax as well as over infiltrated lung tissue and cavities.

A metallic sound of the voice is heard over large cavities, and pneumothorax.

Succussio Hippocratis, or a metallic splashing, is heard when air and fluid are present in the pleural cavity at the same time, if the patient be taken by the shoulders and shaken, as in sero- and pyopneumothorax. A dull or cooing sound is also occasionally heard over phthisical or bronchiectatic cavities.

Pleuritic friction is heard when the pleural surfaces which, in a normal condition are smooth and shining, are roughened by fibrinous deposits, tubercular eruption, or abnormal dryness, and the pulmonary surface of the pleura rubs against the parietal surface during respiration. On adhesion of both pleural surfaces no friction sound is heard. The friction sound is generally of a jerking character, and is either soft or creaking.

It is closely connected with respiration and ceases on holding the breath. It is distinguished from dry râles by being less regular, is not influenced by coughing, and is increased by pressure on the intercostal spaces. Further it appears to be more superficial and nearer the ear. It is strengthened by a strong inspiration. The pleuritic friction is often to be felt on palpation.

The *vocal fremitus*, or pectoral fremitus, is the vibration of the voice transmitted through the bronchi and lung tissue to the chest wall.   It is felt by laying both hands symmetrically on the chest wall while the patient speaks. The strength of the vocal fremitus is dependent upon the strength and depth of the voice, and upon the amount of resistance.

Increase of vocal fremitus occurs in infiltration and compression of a circumscribed portion of the lung when the bronchus leading to this part is unobstructed.   It is noticed in pneumonia, above pleural exudations, as well as over cavities with consolidated walls.

Decrease and absence of the vocal fremitus occurs in very weak or absent voice, in obstruction or stenosis of the bronchus leading to that part where the chest wall is unusually thick, and where there is a *layer of air or fluid* between the lung and chest wall, principally over pleuritic exudations and pneumothorax.   Still, in pleuritic adhesions between the lung and the chest wall, the vocal fremitus within the region of a pleuritic exudation or a pneumothorax may be partly present, or even increased.

CHAPTER IV.

THE SPUTUM.

THE sputum consists not only of the secretion of the tracheal and bronchial mucous membrane, as well as of the pus from the cavernous portions of the respiratory tracts, but also it consists of the secretions of the pharynx and nasal cavity, as far as this is expectorated ; also it consists of the saliva and the secretion of the mucous membrane of the mouth. The remains of food are often mixed with the sputum.

According to the *principal constituents* of the sputum, it is divided into—

1. *Mucous,*
2. *Purulent,*
3. *Serous,*
4. *Bloody,*

and the combined kinds, or *muco-purulent* (principally mucous), *purulo-mucous* (principally pus), *sanguineo-mucous, sanguineo-serous,* etc.

We also distinguish according as the different constituents are *intimately mixed* with the mucus or not.

Pure mucous sputa are found principally in incipient bronchitis. The sputa also of the vault of the pharynx are very thick and often consist of dried mucous masses.

Pure purulent sputa are found in rupture of abscesses of the lung or neighboring organs, or from empyema of the bronchi.

Serous foamy sputa are observed in œdema of the lungs.

Sanguineo-mucous sputa, intimately mixed (brick-red to rust-

colored), are found in pneumonia, hemorrhagic infarct, and carcinoma of the lung (raspberry-jelly sputa) ; *sanguineo-serous* sputa
(prune-juice sputa) are seen in œdema of the lung in the course of
croupous pneumonia. This latter is not to be confounded with the
blood-colored sputa (brown-red, with a stale smell) which are often
expectorated by malingerers and hysterical persons.

Muco-purulent sputa, intimately mixed, are found in diffuse
bronchitis and broncho-blennorrhœa, and in the latter disease the
sputum divides in the spit-cup into three layers. In phthisis pulmonum the sputum is generally purulo-mucous and not mixed, while
the pus is in streaks or balls, or nummular, and surrounded by
mucus. In very large cavities the sputum may run together and be
mixed.

Pure bloody sputa (hæmoptysis) occur when a blood-vessel, or even
a small aneurism in the neighborhood of the respiratory organs, is
eaten through by ulcerations. The blood, in this case, differs from
the blood in hemorrhage from the stomach principally by its being
bright red and not mixed with food.

The *consistency* of the sputa is dependent upon the
amount of mucus in it, and not upon the solid substances.

Smell. A foul smell is caused by decomposition in the
bronchi and the lungs (fœtid bronchitis, gangrene of the
lung)

Color. Apart from the yellow-greenish color caused
by the presence of pus, we may have *red, brown,* or *yellow-greenish* color due to more or less changed condition
of the blood-coloring substance, as in hæmoptysis, infarct
of the lungs, pneumonia, etc.

A *yellow-ochre color* of the sputum is observed from the
presence of hæmatoidin, as in abscess of the lung.

A *green color* may be caused by the coloring matter of
the gall (*e. g.,* pneumonia with icterus), by micro-organisms causing decomposition.

Blue-colored sputa are seen in workmen in dyeworks ;

black sputa in those who inhale much coal dust or soot ; also in iron-workers. In the latter we find occasionally *ochre-colored* and *red* sputa.

Sputa, the *color of the yolk of an egg*, are seen in consequence of bacteria.

The *reaction* of the sputa is generally alkaline.

The *amount* varies according to the cause. Especially large quantities are observed in broncho-blennorrhœa, in large bronchiectatic and tuberculous cavities, and in œdema of the lung.

Morphological Constituents.

Leucocytes are observed as a constancy in sputum, and they are the more abundant, the more the sputum is of a purulent character. They are often in process of disintegration if they are old or if decomposition has set in, as in fœtid bronchitis, gangrene of the lung, bursting of empyema, etc.

Pavement epithelium is from the mouth cavity and from the true vocal cords.

Cylindrical epithelium may come from the nasal cavity, from the upper part of the pharynx, the larynx, and bronchi. They are seldom observed in the sputum, and then only in acute catarrh of the mucous membranes just mentioned.

Alveolar epithelium (whose source in the alveoli of the lungs and in the finest bronchi has not been surely proved) are the large round or oval cells with vesicular nucleus ; and in the protoplasm are often seen particles of fat, of coal, and myeline. These cells are occasionally colored yellow-brown by the coloring matter of the blood (in brown induration of the lung and infarct). They are of no diagnostic importance.

Red blood corpuscles :

Bronchial casts of fibrin are expectorated in fibrinous bronchitis and croupous pneumonia.

Curschmann's corkscrew spiral threads with a bright axis-cylinder are observed in capillary bronchitis, and in bronchial asthma. They may generally be recognized with the unaided eye as fine threads, and are often seen in small sago-like clumps of mucus.

Elastic fibres are observed in the sputum in all destructive diseases of the air passages, and especially in phthisis pulmonum and abscess of the lungs. In gangrene of the lungs the elastic fibres are absent, because there is in this disease a ferment present which dissolves the fibres· In order to show the elastic fibres, it is generally sufficient to mix a suspicious part of the sputum on a slide with a drop of a 10 % solution of caustic potash, and then to examine it. Also a larger amount of sputum may be heated with an equal amount of a 10 % caustic potash solution, and left standing in a glass with sloping sides until a sediment forms, which may be examined under the microscope. Elastic fibres occasionally come from the food.

Shreds of the parenchyma of the lung are especially noticeable in abscess and gangrene of the lung.

Fatty acid crystals and fine curved colorless crystals are found in putrid bronchitis, abscess and gangrene of the lung. They melt to fat on heating the slide. They are found most frequently in yellow-white foul-smelling clumps, as large as a pin's head and larger.

Hæmatoidin appears in amorphous yellow-brownish grains, or in rhombic plates and undulating needle-like crystals of the same color. It is seen in old blood extravasations, and in the lungs, and in abscess of the lungs and neighboring organs.

The Charcot-Leyden crystals are pointed, colorless, shining octahedra, and are especially found in bronchial asthma. They are seen most easily in the yellow fatty flakes and streaks in the sputa.

Crystals of cholesterine, leucine, and *tyrosine* are seen occasionally in abscess of the lung and in putrid expectoration.

Of *animal parasites* are found in the sputum principally echinococcus hydatids and the ova of distomum.

Micro-organisms are present in all sputa, but are especially abundant in putrid decomposition. Of especial importance are the *bacilli* of *tuberculosis* and of *anthrax.* In order to examine for the bacilli tuberculosis, parts are taken from the sputa, free from pus of such a kind as seems to come from a cavity. For directions how to prepare and color preparations, see Chapter XI. The examination for the *coccus of pneumonia* has, as yet, reached no diagnostic importance. Now and then threads of aspergillus (pneumonomycosis aspergillina) are found in the sputum. These are best recognized in a preparation which has been treated with a 10 % caustic potash solution. Also leptothrix threads, which are stained blue by a solution of iodine in iodide of potash, and sarcinæ, as well as the rosettes of actinomycosis are all occasionally found in the sputum.

CHAPTER V.

LARYNGOSCOPY.

THE *larynx* is situated between the upper border of the 3d, and lower border of the 6th cervical vertebra during rest, and rises and falls during respiration, phonation, and deglutition. Very great respiratory excursions are made in stenosis of the larynx, when the head is inclined backward. Very few or no respirations are made in stenosis of the trachea, when the position of the head is inclined forward.

The *percussion* of the larynx gives a tympanitic sound, a higher tone when the mouth is open, and a deeper one when the mouth is closed. The *auscultation* of the larynx and trachea gives loud tubal respiration which is called laryngo-tracheal respiration.

Voice.

We distinguish : (1) An *open* and a *closed nasal voice*, the former when (in paralysis or perforation of the soft palate) the closing of the posterior nares is impossible, and the latter, when the nose is impermeable to air and is obstructed, (polypi, tumors, and stopping of the nose by coryza) ; (2) A *hoarse* voice or one accompanied by disturbing accessory sounds ; (3) A *weak* voice ; and (4) A *want of voice* (aphonia, the voice is without sound) ; (5) *Falsetto voice ;* (6) *Bass* (an unnaturally deep voice in destruction of the vocal cords) ; (7) *Diphthonia ;* and (8) *Tripartite* voice in polypi of the vocal cords.

37

The Muscles of the Larynx.

The larynx is *raised* by the hyo-thyroid, and *drawn down* by the sterno-thyroid ; the *epiglottis* is *raised* by the thyro-epiglottic, and *lowered* by the ary-epiglottic muscles.

The *widening* of the *vocal chink* (abduction of the vocal cords) is carried out by the posterior crico-arytenoid muscle. The same muscle turns the processus vocalis of the arytenoid cartilage outward.

The *closure of the vocal cords* (adduction of the vocal cords) is carried out by the lateral crico-arytenoid muscle, which turns the processus vocalis inward, and by the inter-arytenoid muscle (transverse and oblique), which draws the base of the arytenoid cartilages to each other.

The *tension* of the vocal cords is maintained by the crico-thyroid, which, by fixation of the cricoid cartilage, moves the thyroid cartilage forwards and upwards. Further, it is caused by the thyro-arytenoid muscles, the actual muscles of the vocal cords.

Nerves of the Larynx.

These spring from the vagus, and the motor branches are very likely originally from the accessorius. The *superior laryngeal nerve* supplies the crico-thyroid muscle with motor branches by its external branch ; by its internal branch, the muscles of the epiglottis ; with sensory fibres the mucous membrane of the larynx. The *inferior laryngeal nerve* (recurrens nervi vagi) on the right side curves backward around the subclavian artery, on the left side around the arch of the aorta, goes upward between the trachea and œsophagus, and supplies all the remaining muscles not supplied by the superior laryngeal nerve.

According to recent investigations of Exner [1] there exists, besides the superior and inferior laryngeal nerves, a median laryngeal nerve, which springs from the plexus pharyngeus. The motor and sensory division of the nerves is less simple than the above description.

Paralysis of the Vocal Cords.

In paralysis of the *posterior crico-arytenoid muscle*, the vocal cord cannot be moved outwards in respiration. The paralyzed vocal cord remains, during respiration, near the median line. In paralysis of both cords (Fig. 19, *a*) there is a small crack only between them, and there arises inspiratory dyspnœa. This same thing occurs in *spasm* and *contrac-*

Fig. 19.

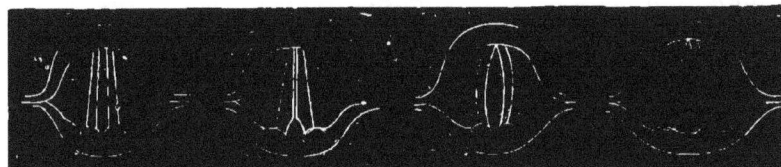

a.	*b.*	*c.*	*d.*
Paralysis of the m. crico-arytænoideus posticus. Position of inspiration.	Paralysis of the inter-arytænoideus. Phonation.	Paralysis of the thyro-arytænoideus. Phonation.	Paralysis of the recurrent laryngeal on both sides. Respiration and phonation.

tion of the *adductor muscles* (lateral crico-arytenoid and inter-arytenoid).

In paralysis of the *inter-arytenoid*, the arytenoid cartilages may approach each other with their processus vocales (lateral crico-arytenoid), but not with bases, and therefore in phonation there remains in the posterior third of the glottis (glottis respiratoria) an open triangle (Fig. 19, *b.*)

In paralysis of the *thyro-arytenoid*, the tension of the vocal cord on phonation is incomplete, and the chord

---

[1] Sitzungsber. der Kaiserl. Akad. d. Wissensch., 89, 1 u. 2. Wien, 1884.

bowed outward with its free edge concave (Fig. 19, c.). When, in addition to this, there is paralysis of the *inter-ary-*

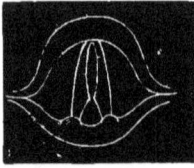

e.
Paralysis of the Mm. thyro-arytænoidei and in-ter-arytænoidei.

tenoid muscles, the chink remains open, and the processus vocales in front are bowed out. (Fig. 19, e.)

In paralysis of the *adductors* (lateral crico-arytenoid and inter-arytenoid), the glottis remains open as a large triangle on phonation (Fig. 19 d.) In paralysis of the lateral crico-arytenoid alone, the glottis has a lozenge shape.

In double-sided paralysis of the *recurrent nerve*, both vocal cords are immovable in the halfway position in phonation as well as in respiration (Fig. 19, d.), the position after death. In paralysis of this nerve on one side, the healthy vocal cord moves in respiration outward normally, and in phonation it approaches the paralyzed cord by crossing of the arytenoid cartilages.

In paralysis of the *crico-thyroid* the vocal cord paralyzed is deeper than the healthy cord on phonation. Also in paralysis of the *superior laryngeal nerve*, there is immobility of the epiglottis on that side as well as anæsthesia of the mucous membrane of the epiglottis (absence of reflex, swallowing the wrong way).

CHAPTER VI.

CIRCULATORY SYSTEM.

Inspection and Palpation.

THE *apex beat of the heart.* The 5th intercostal space on the left side between the parasternal and mammary line, is where the apex beat of the heart is to be found in healthy individuals. In children it often lies in the 4th intercostal space and more outward, and in old people it lies in the 6th intercostal space. The apex beat is lower during a deep inspiration, and lies more to the left, when the individual lies on the left side.

The apex beat is *lower* down in hypertrophy of the left ventricle, aneurism of the aorta, and when the diaphragm is lower, as in emphysema and pneumothorax.

The apex beat is *high* when the diaphragm is pressed upward by abdominal tumors, ascites, tympanites, and contraction of the left lung.

Movement of the *apex beat* and of the *cardiac dulness to the right* is observed in pleural exudation of the left side, and in pneumothorax or in contraction of the right lung. The apex beat lies more to the left in hypertrophy and dilatation of the heart, in collections of fluid and air in the pericardium, and when the mediastinum is pressed to the left.

The apex beat may be of normal strength, or weakened (or even absent), or increased, that is :

(*a*) Simply strengthened.
(*b*) Shaking.
(*c*) Heaving.

41

A *weakening* of the apex beat is observed when the heart is unable to do its work (degeneration of the heart muscle, fatty heart) and when the heart is pressed backward from the chest wall by air or fluid in the pericardium or by an emphysematous lung.

Increased strength of the apex beat is observed in increased activity of the heart (fever, mental excitement, exercise) and hypertrophy of the left ventricle. In the latter case, the apex beat is below and outward, whereas in hypertrophy of the right ventricle, the apex beat is *extended* toward the right side.

Cardiac movement, visible to a greater extent, is seen in materially increased heart action, and when the heart is in contact with the chest-wall to a greater extent, as in contraction of the left lung.

In advanced hypertrophy and dilatation of the heart, and in pericarditis, the *cardiac region projects forward* (voussure).

A *sinking in* of the chest at the apex during the systole denotes adhesions of the heart with the pericardium.

Pulsation in the epigastrium occurs in hypertrophy of the right ventricle, and when the diaphragm is low.

Pulsation of the ascending aorta in the second right intercostal space, as well as of the pulmonary artery in the second left intercostal space, is observed in enlargement (aneurism) of these vessels, as well as in thickening of the borders of the lungs. When the closure of the valve of the pulmonalis may be felt on palpation, it is regarded as pathological, and is caused by stasis in the lesser circulation.

Also in the heart region, *pericardial* friction and a *systolic* or *diastolic* buzzing sound may be felt at all the valves, and the latter has the same importance as the murmur to which it corresponds.

Pulsation in the *bulb of the jugular vein* is an important sign in insufficiency of the tricuspid. This pulsation is synchronous with the systole, and shows that the bulbus valves are incapable of closing, on account of the extension into the jugular vein of the pulsation. Venous pulse may be seen in the arm and in the liver, as in the case of a pulsation extending over the whole liver, and especially over the right lobe. A presystolic venous pulse is observed in overfilling of the right heart when the tricuspid valve is intact. An abnormal fulness of all the veins (*cyanosis*) is seen when the heart is barely able to work (valvular disease) or in obstruction in the lesser circulation.

Diastolic collapse of the veins may occur in pericardial adhesion.

A *capillary pulse* is seen in hypertrophy of the left ventricle (especially in insufficiency of the aorta) when the finger is drawn across the forehead.

Percussion of the Heart.

Cardiac dulness. In healthy individuals, the heart dulness begins above, at the lower border of the fourth rib ; the inner boundary is on the left border of the sternum, the outer boundary is formed by a line drawn from the fourth costal cartilage, curving convexly around, and ending at the apex beat. The inner and under side of the heart dulness measures 5–6 cm. [2–2½ inches]. In children, the heart dulness is, relatively, somewhat greater, and in the aged, smaller. On deep respiration the heart dulness is decreased in size except in adhesions of the left lung. When the patient is on the left side, the heart dulness lies more outward. Increase of the heart dulness occurs in hypertrophy and dilatation of the heart.

The *dulness* is *increased* from above, downwards and outwards in hypertophy and dilation of the left ventricle, while in hypertrophy and dilatation of the right ventricle the heart dulness is broader and lies over the right side of the heart.

Hypertrophy of the *left* ventricle is observed in insufficiency of the aortic and mitral valve, in stenosis of the aortic valve (without dilatation), in aneurism of the aorta and atheroma of the arteries, in nephritis, and after long-continued and excessive bodily exertion,

Hypertrophy of the *right* ventricle begins in an overloading or in obstruction in the pulmonary circulation (mitral insufficiency and stenosis, emphysema, contraction of the lung, defects in the pulmonary valves, and in insufficiency of the tricuspid).

Increase of heart dulness in length and breadth occurs when the whole heart is hypertrophied, or when there is an effusion into the pleural cavity. In the latter case, the dulness is in the form of an equilateral triangle whose apex lies in the 3d–1st intercostal space, and, which lies on the right, beyond the right sternal border, and on the left, beyond the apex beat.

There is also increase of the heart dulness in retraction of the left lung, chlorosis, and in fatty degeneration of the heart, and when the heart is pressed against the anterior chest-wall by pressure upward of the diaphragm (by pregnancy, by tumors of the mediastinum, etc.).

An *apparent enlargement* of the heart dulness occurs in infiltration of the left lung, and in a pleural exudation of the left side,

A *decrease in area of the heart dulness* occurs in atrophy of the heart and in *emphysema*. On presence of air in the pericardium, there is observed, when the patient lies on the back, instead of heart dulness a tympanitic or metallic sound, which changes its seat as the patient moves.

When the ductus arteriosus Botalli remains open, the heart dulness is in the shape of a small quadrilateral figure,

In case of situs viscerum transversus, the heart dulness and apex beat are found in the corresponding place on the opposite side.

Aneurisms of the ascending aorta cause dulness and pulsation at the second and third sterno-costal articulation of the right side. Aneurisms of the arch and of the pulmonary artery cause the corresponding appearances on the left side.

Auscultation of the Heart.

Six sounds are heard over the heart. A systolic sound from each venous opening (mitral and tricuspid valves), a systolic and diastolic tone from the arterial openings (aorta and pulmonary). The systolic sound begins with the contraction of the ventricle, the diastolic with the beginning of the relaxation of the heart, and the closing of the aorta and pulmonary valves.

The *mitral* is auscultated over the apex beat ; the *tricuspid*, at the right sternal border at the fifth and sixth costal cartilage ; the *aortic*, at the second intercostal space, on the right side near the sternum ; and the *pulmonary* valves in the second intercostal space on the left side, very near the sternum.

Two sounds are heard at each opening. The second sound, at the mitral and tricuspid, is transmitted from the aorta and pulmonalis. Over the ventricles the accent is on the first sound (trochæus), over the great vessels it is on the second sound (iambus). The second aortic sound is normally as strong as or even stronger than the second pulmonary sound.

A *strengthened first sound* is heard when the work of the heart is increased, in hypertrophy of the ventricle in chlorosis, in mitral stenosis, also in fever.

A *weakened first sound* occurs in weak conditions, in degeneration of the heart musculature, and in emphysema.

Strengthening of the *second aortic sound*, occurs on increased arterial pressure (*e. g.*, nephritis), and in atheroma of the aorta.

A *strengthened second pulmonary tone* is heard in over-distension of the cavities, in obstruction in the lesser circulation (mitral insufficiency, stenosis and emphysema, and cirrhosis of the lung). In mitral troubles, the strengthened second pulmonary sound is wanting as soon as the tricuspid insufficiency begins.

Metallic-sounding heart tones are heard at times over lung cavities, in pneumopericardium, and dilatation of the stomach.

Reduplication of the heart sounds is occasionally observed in health (depending upon respiration) in unequal tension of the column of blood in both ventricles, which causes the valves on both sides to close at different times.

A *reduplication of the first sound* is observed in hypertrophy of the left ventricle, especially as a consequenc of granular atrophy of the kidneys. A reduplication of the second sound at the arterial openings is due, among other reasons, to mitral stenosis.

The Heart Murmurs.

Systolic murmurs are those heard from the beginning of the first heart sounds to the beginning of the second heart sounds, and all murmnrs from this point, to the first sound again, are called *diastolic*. A diastolic murmur which is heard immediately before the beginning of the next systolic, is called a *presystolic* murmur. The murmurs are heard *after* the sound or *instead* of the normal heart sound. These murmurs may be confined to one heart phase, or they may continue from the one to the next. The character of the murmur may be breathing, blowing, rasping, gushing (a diastolic murmur in aortic insufficiency), groaning, etc.

The *strength of a murmur* is in proportion to the swift-

ness of the blood current, and to the amount of narrowing of the walls, and to their smoothness or roughness.
The murmurs are transmitted most easily in the direction
of the blood current which causes them ; therefore, in
mitral insufficiency, a systolic murmur is heard most
distinctly at the second costal cartilage of the left side,
where the enlarged left auricular appendix lies near the
pulmonary artery, close to the chest-wall.

A systolic murmur of the mitral and tricuspid valves is
indicative of an insufficiency of these valves, and a murmur at the aortic and pulmonary valves, indicates a
stenosis of these valves.

A *diastolic murmur* at the mitral and tricuspid valves
is a sign of stenosis of these valves, and a diastolic murmur at the aortic and pulmonary means an insufficiency
of these valves. Diastolic murmurs are generally of
greater diagnostic importance than systolic murmurs,
and we therefore lay more stress upon these, in the
study of valvular affections.

In heart murmurs, we make a distinction between
pericardial and *endocardial* murmurs, and we divide the
endocardial again into *organic*, and *accidental* or inorganic. The accidental murmurs are generally soft and
blowing, and most frequently systolic, and very seldom
diastolic. They are observed in faulty nutrition, and in
contraction of the heart musculature in high fever and in
changes in the blood (anæmia, chlorosis, hydræmia).
In progressive pernicious anæmia, and bad cases of
chlorosis, accidental diastolic murmurs are also heard.

Pericardial friction murmurs are caused by roughness
of the pericardium, as in deposits of fibrin, tubercle, and
carcinoma, pericarditis, and abnormal dryness of the
pericardium. They are generally slight rubbing, jerk-

ing sounds, and seem very near the ear. They are not always synchronous with the systole or diastole, but are often between both. They are influenced in their intensity by change of position, and by deep inspiration. The normal heart sounds, or endocardial murmurs, may occur with them.

Extrapericardial friction sounds, caused by friction between the pericardium and the pleura, are dependent upon the respiratory movements of the lungs as well as upon the heart's action. They generally cease on holding the breath.

Auscultation of the Blood Vessels.

Over the carotid and subclavian two sounds are heard with each movement of the heart. The first, corresponds to the systole of the heart and expansion (diastole) of the arteries ; the second, to the diastole of the heart (closure of the aortic valve) and to contraction (systole) of the arteries. The first tone arises from distension of the arterial wall, the second is the sound, of the aortic valves transmitted along the vessels. The second sound is absent at the aorta and subclavian in insufficiency of the aortic valves. In stenosis of the aortic and sometimes in insufficiency of the aortic and mitral valves, in atheroma of the aorta, in fever, etc., there is heard *over the carotid* a murmur which is synchronous with the systole of the heart and is called an arterial diastolic murmur.

In the more distant arteries (crural, brachial, radial) no sounds or murmurs are normally heard. On pressure, however, with the stethoscope upon the vessel, there is heard an arterial diastolic murmur, and by still harder pressure, there is a sound called pressure-tone heard.

Abnormal sounds in the small arteries (cubital, palmar, etc.), are heard in aortic insufficiency ; *arterial murmurs* are heard in insufficiency of the aortic valve and in aneurisms. A *double sound* is heard over the *crural artery* in aortic insufficiency, in lead poisoning, and pregnancy.

We auscultate the carotid at the point of insertion of the sterno-mastoid muscle into the clavicle and sternum, or at the inner border of this muscle, on a level with the thyroid cartilage. We auscultate the subclavian in the groove of Mohrenheim, or at the outer part of the supraclavicular fossa.

The cubital and crural arteries are auscultated respectively in the bend of the arm and in the popliteal space. The stethoscope should not be pressed down at all.

In small children from three months on, to the time when the greater fontanelle closes, a systolic blowing murmur may be heard at that point.

A placental murmur is heard in the second half of pregnancy.

When the jugular vein is partly distended (in all forms of anæmia, especially in chlorosis), a whistling, continuous murmur (bruit de diable) is heard over this vein, at the outer border of the sterno-mastoid muscle. This murmur is increased in intensity when the head is turned to the other side. In severe cases of anæmia, a murmur is also heard over the crural vein.

CHAPTER VII.

THE PULSE.

The following varieties of pulse are distinguished.

(1) *The frequency* of the pulse in healthy adults is on an average of 70 beats a minute (60–80), in children 100–140, and in old men 70–90, sometimes more.

The *slower movement* of the pulse (pulsus rarus) is observed in icterus (from the effect of the gall acids on the heart ganglia), in irritation of the vagus, in paralysis of the heart ganglia and of the cervical sympathetic, in increased cranial pressure (first stage of basilar meningitis), in anæmia, in many diseased conditions of the heart musculature (fatty heart), and in stenosis of the aortic valves.

Acceleration of the pulse (pulsus frequens) is observed normally on muscular exertion (to 140), at times in weakened individuals, and after taking food, and pathologically in fever, so that for every 1° [1.8° F.] of increased temperature, there are about 8 pulse beats ; further, in paralysis of the vagus, and in excessively increased cerebral pressure (in the last stage of basilar meningitis), in many neuroses of the heart (stenocardia, morbus Basedowii, etc.), in acute disease of the heart (endocarditis, pericarditis, and many cases of myocarditis), in almost all valvular troubles when there is disturbed compensation, and in collapse.

(2) *Rhythm* (Pulsus regularis and irregularis).

Irregular, arhythmic pulse is observed in extreme old age, and in many diseases of the heart and brain. An irregular pulse is of minor importance in advanced age, whereas in childhood it is observed in severe diseases only, as in basilar meningitis. By pulsus *alternans* is understood that kind of pulse in which a low pulse wave follows a high one. By pulsus *bigeminus* and *trigeminus* is understood a pulse in which there is a long pause after every two or three beats. Pulsus *paradoxus* occurs when the pulse is smaller with each inspiration or may even entirely disappear, as in inflammation of the mediastinum, pericardial adhesions, and stenosis of the air passages. *Retardation* or *unequal size* of the pulse, is where there is a difference between the radial pulse of both sides, or between the upper and lower half of the body, and is observed principally in aneurism and narrowing of the arteries.

(3) *Quickness* (pulsus celer or tardus, fast or slow pulse), *i. e.*, the quickness with which the arteries distend and contract. The sphygmographic curve of the pulsus celer is steep and pointed, that of the pulsus tardus, long and shallow ; the tracing shows a rounded top to the pulse wave.

Pulsus celer is observed in strengthened heart action, and in hypertrophy of the left ventricle, as, for example, in Basedow's disease and in granular nephritis, and is most marked in *insufficiency of the aortic valves* (Fig. 21, No. VIII.). *Pulsus tardus* is observed in extreme old age (*senile pulse*, with rounded top, Fig. 20, No. II.), and in *aortic stenosis.* (Here the tracing shows a wave whose course is long drawn out.)

(4) *Size* (pulsus magnus or parvus), that is, the height of the pulse wave. The pulse is larger in proportion to the impelling power of the heart, and to the amount of blood drawn into the arteries, and the less their tension is. A large pulse is observed in aortic insufficiency, in cardiac hypertrophy, in fever ; and a small pulse in stenosis of the cardiac openings, in cardiac weakness, and in chill.

(5) *Fulness* (pulsus plenus or vacuus), *i. e.*, when the arteries are partly filled only, and a pulsus plenus is observed in increased impelling power of the heart, and when the contractibility and elasticity of the arterial system remain the same, or in hypertrophy of the left

ventricle ; pulsus vacuus is observed when the arterial system is only partly filled with blood.

(6) *Hardness* (pulsus durus or mollis), *i. e.*, the degree of *tension* in the arterial system in proportion to the resistance felt to the finger. A hard pulse is difficult to suppress. A hard wiry pulse is observed in increased impelling power of the heart (hypertrophy of the left ventricle), as well as in a spasm of the arterial muscle (lead colic). A soft pulse is felt in mitral troubles, in fever, and anæmia. An *apparently* hard pulse is observed in atheroma of the arteries.

When the pulse curve is traced with a *sphygmograph*, we notice an ascending and a descending line. Eleva-

Fig. 20.

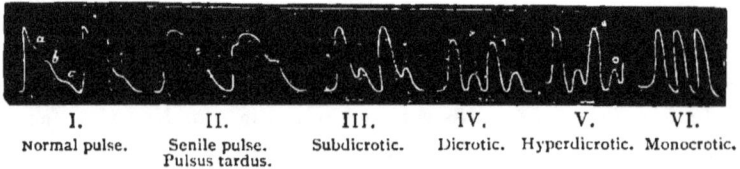

| I. | II. | III. | IV. | V. | VI. |
| Normal pulse. | Senile pulse. Pulsus tardus. | Subdicrotic. | Dicrotic. | Hyperdicrotic. | Monocrotic. |

tions on the ascending part are called anacrotic, on the descending, catacrotic. According to the number of catacrotic elevations, a pulse is called catadicrotic, catatricrotic, etc. In the *normal pulse*, the ascending arm of the curve is straight, while the descending arm shows certain elevations due to the

Fig. 21.

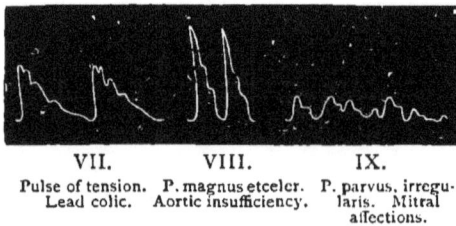

| VII. | VIII. | IX. |
| Pulse of tension. Lead colic. | P. magnus etcelcr. Aortic insufficiency. | P. parvus, irregularis. Mitral affections. |

elasticity of the arterial wall (the *elevation of recoil*, Fig. 20, No. I., *b*), and a greater elevation due to the recoil of the column of blood on the aortic valves, and one or more smaller elevations due to the vibration in the tense *elastic* arterial wall. (Fig. 20, No. I., *a* and *c*). The greater the arterial tension, the more are these elevations of elasticity, and the more distinct is the elevation due to the recoil (lead colic, Fig. 21, No. VII., acute and chronic nephritis). If the arterial tension decreases, the elevation of elasticity disappears, and the elevation of recoil becomes greater and is nearer the beginning of the curve. It may be felt as a recoil wave, and in this case the pulse is said to be *dicrotic.*

A *dicrotic pulse* is observed principally in fever, so that with increasing fever and decreasing arterial tension, the pulse becomes first *subdicrotic* (III.), then *dicrotic* (IV.), *hyperdicrotic* (V.), and finally, with extremely high temperature, it is *monocrotic* (VI.). With a subdicrotic pulse, the elevation of recoil appears before the descending line has reached the base of the curve ; in complete dicrotic pulse, after it has reached the base of the curve ; in hyperdicrotic pulse, the elevation of recoil belongs to the ascending part of the next wave ; and in monocrotic pulse, no elevation of recoil can be recognized.

A slow ascending line, round and broad top, no elevation of elasticity, and a slight recoil, are observed in atheroma of the aorta (pulsus tardus, senile pulse). A slight elevation and long line of descent is observed in stenosis of the aortic valves. *Anacrotic* elevations in the ascending branch of the curve (Fig. 20, II.) occur in diseases of the arterial wall or aortic valves, by the distension of the arteries being slow and jerky.

The *pulse curve in the veins* is the reverse of that in the arteries. This ascends slowly and falls quickly. The ascending branch of the curve is anadicrotic, the second elevation coming from the contraction of the right auricle.

The venous pulse observed in *insufficiency of the tricuspid*, is characterized by its beginning in the diastole, *reaching its maximum in the systole, and continuing through the same.* The venous pulse, however, occurring with a normal tricuspid valve, falls immediately before the beginning of the systole, *i. e.*, before the ascent of the arterial blood wave. To be convinced of this it is only necessary to put the finger on the carotid and follow the venous pulse with the eye. The wave appearing in tricuspid insufficiency, and *outlasting* the arterial pulse, comes from the blood wave, which is thrown back during the systole through its insufficient valve into the auricle and upon the venous system.

CHAPTER VIII.

DIGESTIVE AND ABDOMINAL ORGANS.

The Teeth.

THE *milk teeth* are twenty in number, namely, 2 incisors, 1 canine, and 2 bicuspids on each side of each jaw. The milk teeth come through between the 7th month and the end of the 2d year. The first to appear is the middle inferior incisor in the 6th-8th month. Then follow the remaining six in the 7th-9th month. Then come the upper and lower anterior bicuspids in the 12th-15th month, and at the end of the second year appear the posterior molars. In the 7th year the 2d dentition begins, and the milk teeth fall out in about the same order in which they came.

The permanent teeth are 32 in number, namely, 2 incisors, 1 canine, 2 bicuspids, 3 molar on each side of each jaw. First appears the anterior molar, which comes through behind the temporary bicuspid in the 4th-5th year. In the 7th year, the central incisors follow; in the 8th year, the external incisors, *i. e.*, first the lower and then the upper. In 9th-10th year, the anterior bicuspid appears; in the 10th-11th year, the canine; in the 11th-12th year, the posterior bicuspid. The second (middle) tricuspid appears between the 12th-13th year; the posterior tricuspids, or wisdom teeth, between the 16th and 30th year.

55

The Saliva.

The normal saliva has a specific gravity of 1002 to
1006 ; the normal reaction is alkaline, but it is often
made acid by decomposition in the mouth, as, for example, in diabetes mellitus. The saliva contains only traces
of albumen, and sometimes, but not always, sulphocyan-
ide of potassium (SCNK). This may be recognized by
adding a few drops of hydrochloric acid, and a diluted
solution of chloride of iron, when a blood-red color is
formed, which is taken up on shaking with ether. In
the saliva there is also a diastatic ferment, which may be
shown by adding to a test-tube of saliva a diluted starch
paste, and letting it stand at the temperature of the body.
In a few minutes sugar is formed which may be shown
by Trommer's test.

Œsophagus.

The length of the œsophagus is, in adults, on an average 25 cm. [8½ inches] ; 8 cm. [3¼ inches] below its beginning, it is crossed by the bronchus. The distance
from the upper incisors to the beginning of the œsophagus is about 15 cm. [6 inches]. Accordingly, when the
œsophageal sound passes the distance of 40 cm. [15¾
inches], counting from the upper incisor, we know it is
in the stomach, and if after 23 cm. [9 inches] it comes
upon an obstruction, we may infer that there is a stenosis or diverticulum at the point where the bronchus
crosses the œsophagus. If the sound penetrate 60–70
cm. [25½–27½ inches] and its point can be felt through
the abdominal walls below a line drawn through the anterior superior spine of the ilium, then there is distension of the stomach.

On *auscultating* the œsophagus just to the left of the

vertebral column, there is heard a short murmur just after swallowing, and this murmur in stenosis may be de-
layed, weakened, or may even disappear. On auscultation in the epigastric fossa (or still better, at the angle
behind the left costal curvature and the ensiform cartilage) there is heard, generally immediately after swallowing, a short murmur, or a few seconds later, a longer
murmur (the murmur of Kronecker and Meltzer and the
primary and secondary murmur of Ewald).

Stomach.

Five sixths of the stomach are to the left and one sixth
to the right of the median line. The fundus lies under the
left leaflet of the diaphragm. The lesser curvature of
the stomach and the pylorus are covered by the left lobe
of the liver. The pylorus lies in the right sternal line
about at the height of the tip of the ensiform cartilage.
The greater curvature takes a course about 2–4 cm. [1
inch] above the umbilicus.

In order to mark out the stomach by *percussion*, we
must first determine the position of the diaphragm, and
the borders of the liver and spleen dulness. Between
these organs we come upon the deep tympanitic sound
of the stomach, which may be more or less easily distinguished from the high tympanitic percussion note of
the intestines. The *half-moon shaped space of Traube* is
the upper part of this tympanitic space, which lies between the lungs on one side, and liver and spleen on the
other. Fuller particulars as to the size of the stomach,
are obtained by artificially distending the same with
carbonic acid gas. This is done by administering a
teaspoonful of bicarbonate of soda and tartaric acid, as
separate doses, in water. When the stomach is partly

filled with food, there is found, on percussion in the lower part, a dulness, which changes its position with the movements of the patient. The stomach is considered dilated, when the greater curvature reaches below the umbilical line. There is also a splashing noise heard on shaking the patient.

In palpation, attention should be directed to circumscribed parts which are painful on pressure. They may point to an ulcer or tumor (carcinoma). Tumors of the stomach, in distinction from those of the liver, do not move vertically during respiration.

Examination of the Stomach's Contents.

Under normal conditions, the stomach digestion of a moderately hearty meal is over in 6–7 hours and the stomach is again empty. If, after this time, lavage of the stomach shows large quantities of food débris, it must be considered as a sign of disturbed digestion.

To accurately define the digesting strength of the stomach, and the amount of acid the gastric juice contains, a sample of the digestive mixture [1] is drawn off with a stomach sound two hours after eating.

Test of the Amount of Acid Contained.—The reaction is first tested with litmus paper. An acid reaction may be caused by hydrochloric acid, or by the organic acids (lactic, butyric, acetic). To test for *hydrochloric acid*, a few drops of a diluted watery solution of methylviolet or tropæoline are added to a few drops of filtered gastric juice. Even 0.1–0.2 ‰ of hydrochloric acid causes a distinct colored precipitate of blue or reddish brown, while a much greater concentration is required to produce the same precipitate in the case of the

---

[1] According to Leube, the gastric-juice secretion is excited by pouring 100 ccm. [6 ounces] of ice-water into the stomach after it has been washed out with tepid water, and ten minutes later syphoning off a sample.

other acids. To test for *lactic* acid, a few drops of the filtered gastric juice are added drop by drop to 1–2 drops of a reagent consisting of 3 drops of the solution of the sesquichloride of iron, 10 cm. [2½ drachms] of a 4 ℀ carbolic-acid solution, and 20 cm. [5 drachms] of water. The original amethyst blue is turned yellow by lactic acid, and pale gray by hydrochloric acid and butyric acid (Uffelmann). When lactic and hydrochloric acids are both present, the former is removed with ether, and the residue is tested for the latter. All these tests are not very reliable, and are influenced by the degree of the acidity and the presence of such bodies as the peptones, neutral salts, etc., preventing the reaction. Still, it can generally be presumed that when the reaction shows a large amount of hydrochloric acid present, a carcinoma of the stomach is probably to be excluded.

In order to test the *digesting strength of the gastric juice*, to two test-tubes containing gastric juice a bit of washed fibrin, and to one of these test-tubes a few drops of 1 ℀ hydrochloric acid, are added, and both tubes are put into the incubator at body temperature. If after 6–12 hours the fibrin in neither tube is dissolved, there is evidently want of pepsin in the sample, and if the fibrin in the gastric juice containing the 1 ℀ hydrochloric acid is alone digested, then we surmise that this gastric juice contains pepsin, but no hydrochloric acid. When the gastric juice is normal, the fibrin in both tubes should disappear in 1–2 hours.

Vomited matter may contain :

Mucus (in gastric catarrh),

Swallowed saliva (in the morning sickness of drunkards). This may be recognized by its containing ferrocyanide of potash (showing a blood-red color on addition of a solution of chloride of iron).

Blood (ulcer and carcinoma of the stomach, cirrhosis of the liver). This may be either unchanged, or is digested to a brown, coffee-ground mass, due to its long stay in the stomach. In this latter case, the red blood-corpuscles are dissolved, and the hæmoglobin changed to hæmatin, which may be shown with the hæmin test (*vid.* page 1).

Gall is also found when the vomiting is frequent and of long duration. In uræmia, *urea* and *carbonate of ammonia* are found in the vomitus.

Remains of food, which may be more or less altered, either by the process of digestion or by micro-organisms. The effects of the latter causing fermentation and decomposition, form lactic, butyric, and acetic acids out of the carbohydrates (starch and sugar) ; free fatty acids out of the neutral fats ; peptones, leucin, tyrosin, phenolindol, skatol, sulphuretted hydrogen, and ammonia out of the albuminous substances. These latter are products of advanced decomposition, and are found principally when the contents of the small intestines regurgitate into the stomach and are vomited (vomiting of fæces).

On *microscopical* examination of the vomitus, there are found apart from the remains of food (cross-striped muscular fibres, fat, starch, vegetable substances, etc.) pavement epithelium of the mouth and œsophagus, and more seldom cylindrical epithelium of the stomach, as constant leucocytes, schizomycites of the most varied kind ; sometimes yeast-cells, sarcinæ, and oidium albicans.

Liver.

The upper border of the *liver dulness* begins at the lower border of the right lung and of the heart. The lower border is, in healthy individuals, in the axillary line between the tenth and eleventh ribs, at the curvature of the ribs in the mammary line ; and at the median line it lies between the xiphoid process and the umbilicus ; it then takes a curved direction upwards and reaches the diaphragm and generally the heart apex between the parasternal and mammary line. In deep inspiration, when the patient lies on the left side, the liver dulness is smaller, because the lung border comes down further. The lower border of the liver moves up and down slightly during respiration.

The *liver* is *forced down* in emphysema, pneumothorax, pleurisy, and pericardial exudations.

It is *pushed upwards* in contraction of the right lung and in increased pressure from below in the abdominal cavity, as in peritonitis, ascites, tumors, and pregnancy. From this cause the anterior edge of the liver may be turned up, causing a material decrease in the size of the liver-dulness.

Hypertrophy of the liver occurs in the first stage of cirrhosis, in congestion, in fatty and waxy liver. Decrease in the size of the liver-dulness occurs in atrophic nutmeg liver, acute yellow atrophy, in the second stage of cirrhosis, and when the transverse colon lies between the abdominal wall and the liver. When air enters the peritoneal cavity, there is complete absence of liver-dulness in the median line.

In healthy adults the surface and edge of the liver cannot be felt. It is, however, resistent and to be felt in congestion, cirrhosis, amyloid degeneration, multilocular echinococcus, and not so easy to be felt in fatty liver. Inequalities of the liver-surface and tumors may be easily felt in cirrhosis, syphilis of the liver, abscess, carcinoma, and echinococcus. When the echinococcus cysts are present there is a slight fluctuation to be felt over the liver (hydatid purring).

In the *liver of tight lacing* that part of the right lobe is felt as a round tumor below the ribs and separated from the rest of the liver by a horizontal furrow. In *wandering liver*, the organ is dislocated downwards (also in the upright position) and abnormally movable.

The enlarged *gall bladder* can sometimes be percussed about 5 cm. [2 inches] to the right of the median line at the lower border of the liver and can be felt as a round tumor.

The Spleen.

The normal *splenic dulness* is in the left hypochondrium, between the ninth and eleventh ribs and reaches forward to the costo-articular line (drawn from the left sterno-clavicular articulation to the tip of the eleventh rib) and backward to the spinal column. The height (breadth) of the spleen-dulness is, in the middle axillary line, 5–6 cm. (2–2½ inches). On inspiration and when lying on the right side, the splenic dulness is made smaller by the lower border of the left lung moving down.

The *spleen* is *lower down* in a pleural exudation of the left side, in pneumothorax, and emphysema. In ascites, tympanites, tumors of the abdomen, the spleen is pressed upward against the diaphragm and the dulness is made smaller.

Hypertrophy of the spleen is observed in almost all infectious diseases (typhoid fever, typhus, pyæmia, the acute exanthematous diseases, beginning secondary syphilis, in many forms of pneumonia, etc.) ; further in leucæmia, amyloid degeneration, cirrhosis of the liver, hæmorrhagic infarct of the spleen, echinococcus, and in the severer forms of intermittent fever. When the spleen is greatly hypertrophied the end may be felt under the left border of the ribs.

An *apparent hypertrophy* is observed in pleural exudations of the left side and in infiltration of the left lung.

The *splenic dulness* is absent in wandering spleen and on the presence of air in the peritoneum when the patient lies on the right side.

In *situs viscerum transversus* the splenic dulness is on the right side and the hepatic dulness on the left side.

Abdomen.

The abdomen emits under normal circumstances a tympanitic sound of varying height in all parts.

A *sinking in of the abdomen* is observed when the intestines are empty and contracted (in inanition, meningitis, lead colic).

Distention of the abdominal walls occurs in overloading of the stomach and intestines with air (tympanites) or fluid, in collections of air or fluid in the peritoneum and intestines.

In collection of free fluid in the peritoneum, *ascites*

(in heart disease and congestion of the portal system, especially cirrhosis of the liver) the abdomen is laterally distended, and in the middle, flat when the patient is recumbent. The fluid, which may be found by dulness and fluctuation, shows a horizontal boundary line above when the patient stands, and changes its position quickly when the patient moves about.

In inflammatory exudations of the peritoneum, the abdomen is generally equally distended, and the fluid is often encysted and immovable. If the fluid is very great in amount the intestines do not reach the abdominal wall (also when the mesentery is shorter) and there is everywhere dulness on percussion.

When air enters the peritoneum, it always seeks the highest level and causes the liver-dulness, *i. e.*, its central part, or the splenic dulness, to disappear, according to the position of the patient. In tumors there is, according to the position, an unequal distention of the abdomen, *e. g.*, in tumors of the liver and spleen there is a distention in the upper part of the abdomen, and in uterine and ovarian tumors in the middle and lower parts.

In peritonitis there is occasionally a friction sound to be felt and heard.

Addendum.

The Fæces.

The fæces consist of

(1) The remains of the food altered by the processes of digestion and decomposition.

(2) The digestive juices in the intestines, and

(3) Certain products of excretion which come from the body through the glands opening into the intestines, *e. g.*, the salts of the heavy metals, iron, lead, mercury, etc.

As to the *consistence* we distinguish *firm, thick fluid, thin fluid,* and *watery.* The two last kinds are not considered normal unless they are caused by the diet or by purgatives. Watery passages (diarrhœa) appear when the food passes through the intestines so quickly that the absorption is incomplete, or, more rarely, when there is an exudation into the intestines, as in dysentery.

Mucus may form a glassy covering to the fæces, or be mixed in large coarse lumps with it (in affections of the large intestine, even in its lower parts), or it is intimately mixed in small particles with the fæces (in disease of the upper part of the large or small intestine). If the mucus is colored with gall, or if it reacts with the test of Gmelin, we may consider that the small intestine is affected. Purulent mucus is observed in ulceration of the intestines. Large cylindrical masses of mucus are passed in the so-called mucous colic (Nothnagel).

The *color* of the fæces is caused principally by the coloring substance of the gall. Generally the coloring substance is altered by bacteria and reduced to hydrobilirubin, but sometimes the coloring substance of the gall appears unchanged, as in the yellow and green passages of infants, and when the peristalsis is especially rapid.

If the gall be absent in the intestine (as in icterus) the passage contains abundant fat, and appears therefore gray, greasy, and clayey, and on shaking with water there is a peculiar play of colors noticed. The passages show an abundant amount of fat and a similar appearance when the absorption of fat does not take place on account of various diseases of the intestinal mucous membrane or of the chylopoëtic system.

The color, as well as the consistence and amount, of the fæces is also dependent upon the *food taken*. In an almost exclusively *meat diet*, firm, brownish-black fæces in small amount are passed. In a diet of *starchy foods* (bread, potatoes) the passage is yellow-brown, soft, foamy, and in large quantity ; in an exclusively *milk diet,* yellow-white and firm ; in an *egg diet,* soft and white, etc. Further, the color is changed by drugs, *e. g.,* iron and bismuth make it black, forming sulphate of iron and sulphate of bismuth ; mercurial preparations, and especially calomel, greenish brown, making a combination of the coloring matter of the gall with mercury ; rhubarb, yellow-brown, and logwood preparations, reddish brown. *Blood* from the upper intestinal tracts mixed with the fæces makes a tarry, blackish-red color ; but if the blood comes from the lower parts (in dysentery and hæmorrhoidal bleeding), it is generally red and unaltered.

In *typhoid fever* the stools generally have the appearance of a badly cooked pea-soup ; in *cholera* they resemble rice water ; and in *dysentery* they contain bloody mucus.

In the *microscopical examination* of the fæces there are found shreds of cross-striped muscle fibres and of the animal tissues ; further, vegetable substances, such as spiral threads, but rarely starch kernels. Fat appears in the form of drops and of glassy clumps, as well as needle-shaped crystals. The latter point to a disturbed fat absorption, and occur most abundantly when the gall does not flow into the intestine. Also the coffin-top-shaped crystals of the ammonia-phosphate of magnesia, and clumpy crystals of other lime salts, as well as the needle-shaped crystals of Charcot-Neumann, are found in the stools.

Cellular elements are also found, *e. g.,* leucocytes in intestinal catarrh, especially in ulcerations. Red blood corpuscles are rarely seen in intestinal hemorrhage, as they are generally already destroyed. Cylindrical epithelial cells are often found in intestinal catarrh, and are often seen in the process of disappearance. Pavement epithelium in the stools comes from the anus.

Micro-organisms occur in large numbers in the stools. The proof of the presence of the bacillus of tuberculosis and of Asiatic cholera is of diagnostic importance. The presence of the latter can be proved by further culture only.

For the account of the animal parasites see Chapter XI., and for the analysis of the concrements of the gall and fæcal stones see Chapter XIII.

CHAPTER IX.

THE URINE-PRODUCING SYSTEM.

The Genito-Urinary Organs.

THE *kidneys* lie on both sides of the vertebral column from the level of the twelfth dorsal to the first to third lumbar vertebra. The right kidney borders above on the liver, the left on the spleen. The lower and outer borders of the organs are determined by percussion. The outer border is about 10 cm. [4 inches] external to the spinous processes (renal-hepatic and renal-splenic angle). The kidney dulness may be absent in disloca-tion of the kidney (wandering kidney is generally on the right side), or it may be increased by tumors of the kid-ney (hydro-nephrosis, neoplasms and echinococcus of the kidney). The latter may generally be felt by press-ing deeply into the anterior abdominal walls.

The *urinary bladder* when filled may generally be felt and percussed as a round tumor in the median line above the symphysis pubis.

The Urine.

The products of decomposition of fat and of the carbohydrates leave the body essentially as carbonic acid and water by the lungs ; the last products, however, of the decomposition of the albumen pass out almost exclusively in the form of urine. Therefore an ex-amination of the urine gives us information, qualitative and quanti-tative, as to the passing out of the products of decomposition of the albumen.

The *amount* of urine excreted in health by men is about 1,500 to 2,000 ccm. [40–60 ounces], and by women 1,000 to 1,500 ccm. [30–40 ounces] in a day. Under 500 ccm. [15 ounces] and over 3,000 ccm. [90 ounces] is almost always pathological.

A lasting increase in the amount occurs in diabetes mellitus and insipidus, in granular atrophy of the kidney, in pyelitis as well as in absorption of collections of fluid in the body. A decrease of urine occurs in fever, in acute and chronic parenchymatous nephritis, in cholera, in profuse sweating, as well as in the formation of exudations and transudations ; further in valvular heart disease, and in other diseases with lowering of the blood pressure.

The *specific gravity* varies in healthy individuals under ordinary conditions of nourishment, between 1,015 and 1,025. A decrease (1,002) occurs in renal disease and in diabetes insipidus, an increase (1,060) in diabetes mellitus and in fever.

The specific gravity is measured by dipping the dry urinometer into the fluid cooled to the surrounding temperature, and then reading off the amount at the level of the fluid.

If the two last figures of the specific gravity be multiplied by the coefficient of Häser 2.33, the result is approximately the amount (in grams) of solid substances in 1000 ccm. [30 ounces] of urine. From this the amount for the 24 hours may be calculated. Thus if the amount of urine be 2000 ccm. [60 ounces], and the specific gravity 1012, then there are 55.92 grams [about 530 grains] of solid substances in the urine.

$$2.33 \times 12 = 27.96 \times 2 = 55.92.$$

The *reaction* of normal human urine is acid, especially on account of the acid phosphates of sodium (NaH_2PO_4) contained in it.

The *acid reaction* of the urine *is greater* in proportion to its concentration—*e. g.*, after strong perspiration, as well as in increased albuminous metabolism, as in fever.

The *reaction is slightly acid, neutral, or alkaline* in very dilute urine, and after taking the alkalies of carbonic acid or vegetable acid (the latter being reduced in the body to carbonic acid, as in a vegetable diet) ; again, when much hydrochloric acid of the stomach has been removed from the organism by habitual vomiting or by lavage. Also in the rapid absorption of exudations and transudations, the reaction of the urine is less acid ; whereas in the collection of this fluid, the urine is strongly acid.

As soon as the urine becomes neutral or alkaline, the earthy phosphates are precipitated (basic phosphate of lime and magnesia $Ca_3 [PO_4]_2$ and $Mg_3 [PO_4]_2$) ; also occasionally the carbonates of the alkaline earths fall to the bottom as a white flocculent sediment. This is dissolved at once by the addition of acids, but not by heating nor by adding alkalies (which distinguishes it from uric acid sediment). Slightly acid or neutral urine becomes sometimes cloudy on heating, by separating the earthy phosphates. This cloudiness disappears on adding acids, or on cooling off—a thing which distinguishes it from the cloudiness of albumen.

If the *urine be decomposed by the presence of bacteria* in the bladder (cystitis), or after it has been passed, the reaction is also alkaline (the alkaline urine fermentation) from the carbonate of ammonia formed from the urea. The ammoniacally decomposed urine has a bad odor and develops hydrochlorate of ammonia by holding over it a glass rod moistened with hydrochloric acid. While there is a moderate amount only of the crystals of ammonio-magnesian phosphates ($Mg NH_4 PO_4$) in the sediment of non-decomposed alkaline urine, these crystals (coffin-lid crystals) appear in abundance in ammoniacally decomposed urine, which may also contain in addition the thorn-apple-shaped crystals of the urate of ammonia. The blue mark on litmus paper from ammoniacal urine disappears on letting the paper dry in the air, while the blue spot from urine whose alkalinity is due to a fixed alkali remains after the litmus paper is dry. If there is a sediment of pus in the urine, it has a crumbly appearance if the urine is acid ; while in alkaline, decomposed urine, this sediment forms itself into balls and mucous, thick thread-like clumps.

Normal Constitutents of the Urine.

Urea (U), $NH_2 CO NH_2$, is freely soluble in water and alcohol. The daily amount excreted by healthy individuals is between 20–40 grams [300–600 grains] ; it is increased in an albuminous diet and in an increased loss of the albumen of the body—*e. g.*, in diabetes mellitus (to more than 100 grams [1,500 grains]), in fever (to 50 grams [750 grains]), in phosphorous poisoning, and in dyspnœa. It is decreased in inanition (to 9 grams [135 grains]), in a diet poor in nitrogen ; further, in uræmia, and in acute yellow atrophy of the liver.

The urea is changed by bacteria or through the effect of strong alkalies into the carbonate of ammonia, by taking up water, thus :

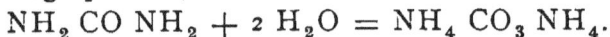

$$NH_2 CO NH_2 + 2 H_2O = NH_4 CO_3 NH_4.$$

If urea be heated dry, biuret is formed, and its watery solution added to caustic potash and a drop of a sulphate of copper solution, gives a violet color (biuret reaction.)

To *show* the *presence of urea* (*e. g.*, in sputa, vomitus, transudations, etc.), the evaporated fluid is extracted with alcohol, filtered, the filtrate evaporated, the residue dissolved in a little water and concentrated nitric acid added to it. After standing in the cold a little while, the hexagonal crystals of nitrate of urea appear.

Since the *nitrogen* which appears in the urine, and which comes from the changes of the albumen in the organism, not only appears as *urea*, but also in other nitrogenous compounds, at least in relatively small amounts, it is better to find out the *entire amount of nitrogen* in the urine, instead of the amount of urea, in order to draw conclusions as to the changes of albumen in the organism. As to the exact methods of determining the nitrogen see the text-books.[1]

In order to find out approximately the amount of nitrogen in the urine, the following modification of the trituration method of Liebig by Pflüger is sufficient :

Fig. 22.

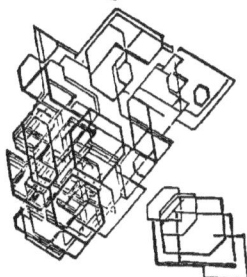

Nitrate of Urea.

A row of thick drops of soda paste (a mixture of soda and water) is dropped on a glass plate which has a black background, then 10 ccm. [2–3 drachms] of urine are measured out with a pipette and dropped into a beaker glass, then let 1 ccm. [15 drops] of a nitrate of mercury solution [2] drop from a graduated burette upon each one of the drops of soda paste. If the point of contact of the two drops continue white, the nitrate of mercury should be added until it turns yellow and until the yellow does not disappear on stirring. Then the amount of the nitrate of mercury used is to be multiplied by 0.04 in order to obtain the percentage of nitrogen in the urine. If, for example, 13 ccm. [½ ounce] of the nitrate of mercury solution be used before the yellow color appears, the result is $13 \times 0.04 = 0.52 \% N$.

Uric Acid, $C_5H_4N_4O_3$.—The daily amount passed is between 0.2 gram [3 grains] and 1.0 gram [15 grains],

---

[1] Neubauer und Vogel, Anleitung zur qualitativen und quantitativen Analyse des Harns ; Salkowsky und Leube, die Lehre vom Harn ; Löbisch, Anleitung zur Harnanalyse ; Hoppe-Seyler, Handbuch der physiologisch- und pathologisch-chemischen Analyse, and many others.

[2] A litre [30–32 ounces] of mercurial nitrate solution contains 71.48 grams [1,070 grains] of mercury. For its preparation, see the text-books just mentioned.

and it generally decreases and increases with the varying
amount of urea.   It is considerably increased in leucæmia.   During an attack of gout it is said to decrease,
and after the attack to increase.

Uric acid is normally present in the urine as neutral
urate of sodium, which is freely soluble in water.   In

Fig. 23.

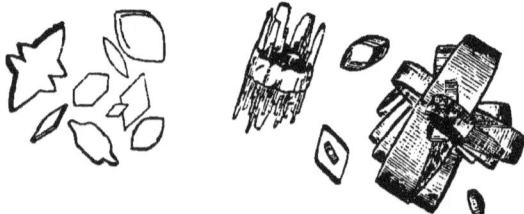

Comb-shaped and Whetstone-shaped Crystals of Uric Acid.

concentrated and strongly acid urine, in fever and after
severe sweating, the *acid urate of sodium* is present after

Fig. 24.

Brick-dust Sediment
of the Urate of Am-
monia.

the urine has stood for some time in the
cold.   This acid urate of sodium is easily
soluble in warm urine, and only with difficulty soluble in the cold.   Red-colored,
brickdust   sediment   is   precipitated   (by
uroerythrin), which is re-dissolved on heating, or on adding caustic potash.   *Urate of ammonia* is
found as thorn-apple-shaped crystals in

Fig. 25.

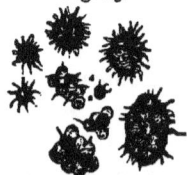

Thorn-apple-shaped
Crystals of the Urate of
Ammonia.

decomposed urine. *Free uric acid*, which
is almost insoluble in water, often appears in strongly acid urine, especially
after long standing.   It forms at the
bottom of the vessel a heavy, hard, red,
crystalline   powder,   which   under   the
microscope shows crystals of the shape of whetstones,
combs, casks, and spear-heads.

In order to test for uric acid, some of the substance (sediment or concrement of the urine) is mixed upon the top of a porcelain crucible with a few drops of nitric acid, and slowly evaporated. An orange-red spot is formed, which turns purple on adding ammonia, and blue on the addition of caustic potash. This is called the *murexide test.*

Oxalic acid, $(COOH)_2$.—The daily amount is as much as 0.02 $[\frac{1}{3}-\frac{1}{4}$ grain], and it appears in the sediment as oxalate of Fig. 26. lime (insoluble in acetic and soluble in hydrochloric acid) in very small, shining, octahædral crystals (envelope shape), or in needle shape.

Envelope-shaped Crystals of the Oxalate of Lime.

Sulpho-cyanide of potash, SCNK.—The daily amount excreted is 0.05 gram $[\frac{1}{12}$ grain]. It gives, with diluted solution of the chloride of iron, a red color, which does not disappear on adding hydrochloric acid, and which is soluble in ether.

Hippuric acid, $C_9H_9NO_3$.—The daily amount excreted is 0.1–1.0 gram $[1\frac{1}{2}$–15 grains]. It is formed in the kidneys by the combination of benzoic acid and glycocoll, and often appears in needle-shaped or rhombic prisms, which are like the triple phosphates, but are insoluble in acetic acid.

Xanthine, $C_5H_4N_4O_2$, and *Hypoxanthine,* $C_5H_4N_4O.$ — The latter is found in the urine in leucæmia.

Creatinine, $C_4H_7N_3O.$—The daily amount, 0.5–1.0 gram $[7$–15 grains], is increased in an abundant meat diet and in increased metabolic muscular change, and decreased in inanition and convalescence.

Aromatic oxi-acids, as paroxyphenol acetic acid, give like the phenol group a red color when treated with Millon's reagent.

The *Phenols : Carbolic acid*, C_6H_5OH, *Hydroquinone,* C_6H_4 $(OH)_2$, *Cressol* $CH_3C_6H_6OH.$—The phenols exist in the urine in the form of the sulphates, as the so-called ethereal sulphates. An increase of this denotes decomposing processes in the organism. To detect the phenols in the urine, to 100 ccm. [3 ounces] of urine 5 ccm. $[1\frac{1}{4}$ drachms] of concentrated sulphuric acid are added, and the whole distilled in a retort. Bromine water is then added to the distillate, and if carbolic acid is present, there is formed a yellow-white precipitate of tri-bromo-phenol.

Indican (indoxyl sulphate of potassium), C_8H_6NO

SO_3K.—Indol, the result of the decomposition of albumen in the intestinal canal, or of putrid suppuration, is re-absorbed and oxydized in the organism to indoxyl, which, in the urine, combines with sulphuric acid and passes out as the indoxylsulphate of potassium (indican). This latter splits up on adding concentrated hydrochloric acid and the chloride of lime (as an oxydizer), and forms indigo blue. Indican is increased by an abundant meat diet ; further, in putrid suppuration, in diseases of the stomach and intestines, in abnormal putrefaction of the ingesta, and is most increased in intestinal obstruction. From the amount of indican in the urine a conclusion may be drawn as to the intensity of the processes of the decomposition of albumen in the intestinal canal.

In order to test for indican, to a small quantity of urine, ¼ of its volume of a 10 % solution of the acetate of lead is added, by which a number of disturbing substances are precipitated, and removed by filtration. To the filtrate is then added an equal part of hydrochloric acid and one or two drops of a concentrated solution of calcium chloride, which is one half diluted with water. The chloride-of-calcium solution is added drop by drop until a blue color appears. Too much chloride of calcium hinders the formation of the indigo. A few ccm. of chloroform are then added, and the whole shaken, which brings out the indigo.

Inorganic Constituents of the Urine.

Hydrochloric acid HCl is present principally combined with sodium as common salt. The amount of sodium chloride in the urine is about one half the amount of urea present, *i. e.*, between 11 and 15 grams [150 and 225 grains]. It depends principally upon the amount of salt taken with the food. It is lessened in inanition and in fever, especially in pneumonia. Indeed, in the latter disease there is often so little present in the urine that the addition of a nitrate of silver solution causes a slight cloudiness only, while normally the chloride of silver is precipitated in large quantities. The chlorides are increased (to as much as 55 grams [825 grains]) during a rapid re-absorption of exudations.

Sulphuric acid H_2SO_4. The daily amount excreted is from 2.0 to 2.5 grams [½–¾ drachm], and it appears partly as ethereal sulphuric acid in combination with phenol, indoxyl, etc., and partly as "preformed" sulphuric acid. The proportion of the former to the latter kind of sulphuric acid is about as 1:10. In carbolic acid poisoning, however, the entire amount of sulphuric acid present may be in combination with carbolic acid. In order to test for the ethereal sulphuric acids the urine is first rendered slightly acid with acetic acid and then barium chloride is added in excess, by which the preformed sulphuric acid alone is precipitated and may then be removed by filtration. The filtrate is then treated with concentrated hydrochloric acid and then heated. By decomposition of the ethereal sulphuric acids a precipitate of barium sulphate is formed from which the amount of ethereal sulphuric acid may be determined.

Phosphoric acid H_3PO_4. The daily amount excreted is 2.5–3.5 grams [¾–1 drachm] of which two thirds are combined with alkalies and one third with alkaline earths (lime and magnesia). The daily amount of the earthy phosphates is 1.2 grams [18 grains].

Carbonic acid CO_2 is present in human urine in very small quantities and is more abundant after taking fruit and vegetable food, after many drugs, as well as in decomposed urine. When large quantities of the carbonates are present the urine effervesces on the addition of acids, and causes a white deposit on a glass rod mois-

tened with baryta water and held over the mouth of the
test tube. Carbonate of lime is present in the sediment
as small spherical and biscuit-shaped [dumb-bell crystals]

Fig. 27.   Fig. 28.   Fig. 29.

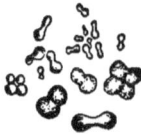

Spherical and Biscuit-shaped [dumb-bell] Crystals of the Carbonate of Lime.

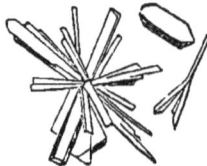

Neutral Phosphates of Lime.

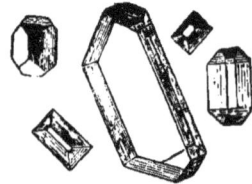

Coffin-lid Crystals of the Ammonio-Phosphate of Magnesia.

bodies which dissolve with the formation of bubbles on
the addition of acids.

Sodium. The daily amount excreted is 4–6 grams
[1–1½ drachms], in form of sodium oxide Na_2O. *Potassium.* The daily amount excreted is 2–3 grams [30–45
grains], in form of potassium oxide K_2O. During fever
the amount of sodium decreases, while the amount of
potassium is 3 to 7 times as great. *Ammonia* NH_3 is
present in unfermented urine in small quantities only,
(0.6–0.8 gram [10–12 grains]). It is much decreased in
many cases of diabetes. *Calcium.* The daily amount
excreted is 0.16 gram [2½ grains], in form of CaO.
Magnesium. The daily amount excreted is 0.23 gram
[3½ grains], in form of MgO.

The *sulphate of calcium* (gypsum) is present in the
sediment in form of fine oblique prisms and needles,
which are not soluble in acetic acid. *Neutral phosphate
of calcium* is present in form of wedge-shaped crystals
which unite to form rosettes. The *ammonio-magnesian
phosphates,* or triple phosphates, occur as shining coffin-
lid-shaped crystals. The two last of these crystals mentioned are soluble in acetic acid.

Iron is present in the organism in combination and therefore appears in the ash of urine only.

Pathological Constituents of the Urine.

Albumen (serum albumen and serum globulin). In order to make use of the following tests, the urine should be clear, and filtered if not clear.

I. *Heat Test.*—The urine is heated to the boiling point in a test-tube, and then one or two drops of diluted acetic acid should be added. Instead of the acetic acid, nitric acid may be used, in which case ten to twenty drops should be added. If the cloudiness caused by heating be dissolved by the acid, then it was not caused by albumen, but by the phosphates and carbonates of lime and magnesia which are freely soluble in the acids. If the cloudiness remain, or if it appear on the addition of acid, it is caused by albumen.

A cloudiness often appears after adding acetic acid to the urine when it is warm, or when it has cooled off. In this case it is not due to mucin, but to albumen.

If the precipitate of albumen be taken from the filter and its volume approximated after three to twelve hours, an approximate result, as to the amount per cent. of albumen in the urine may be obtained. When the amount of albumen is 2 % to 3 %, the whole fluid is completely coagulated. When there is 1 % of albumen present, the coagulum in the test-tube reaches half way up to the level of the urine.

When 0.5 %, $\frac{1}{3}$ the way up.

" 0.25 %, $\frac{1}{4}$ " " "

" 0.1 %, $\frac{1}{10}$ " " "

" 0.05 %, the curved part of the tube is barely filled with albumen, and when there is less than 0.01 % present, there is a slight cloudiness, but no precipitate.

II. *Heller's Test.*—The test-tube containing the urine is held obliquely and concentrated nitric acid is poured slowly down the side of the tube so as to flow below the urine. If albumen be present, there is formed a sharply defined ring-shaped cloudiness at the point of contact between the urine and the acid. Besides albumen, a precipitate in very concentrated urine may be caused by the presence of urea, in which case the ring is higher and not so clear. A cloudiness may also be caused by nitrate of urea, and in this case the precipitate is crystalline and does not appear until after standing a long time. A cloudiness may occur from the resinous substances, as after taking copaiva, styrax, turpentine, etc., but in this case the precipitate is dissolved, after cooling, in alcohol. The ring of albumen may be colored blue or green by indigo, or by the coloring matter of the gall.

III. *Test with Acetic Acid* and *Ferrocyanide of Potassium* in the cold.—If to some urine three to five drops each of acetic acid and a 10 % solution of ferrocyanide of potassium be added, there occurs a precipitate from the presence of *albumen* or *hemialbumose*. If the urine be taken in very small quantities, the precipitate appears only after a few minutes.

IV. *Biuret Test.*—The urine is first to be made alkaline with caustic potash, and then 1–3 drops of a diluted solution of sulphate of copper are to be added, and if *albumen, hemialbumose*, or *peptone* be present, a reddish violet solution is formed.

[V. *Picric Acid Test.*—A delicate and convenient test used long ago in Germany and rediscovered in 1882 by George Johnson. The dry acid may be dissolved in the urine, or a saturated solution may be used, into which the urine should be slowly dropped, and if albumen is present a cloudiness appears at once.]

For the quantitative determination of albumen, see Chapter X.

Hemialbumose (Propeptone) is an intermediate state be tween albumen and peptone. This is not precipitated by heating, but by nitric acid, acetic acid, and ferrocyanide of potassium, as well as by acetic acid and sodium chloride. All these precipitates have the property of *dissolving on heating* and reprecipitating on cooling.

To test for hemialbumose it is necessary, first, to remove the albumen. For this purpose, to the urine (or to any other fluid to be examined, as the contents of the stomach) 5 to 10 drops of acetic acid and ⅛ of its volume of a concentrated salt solution are added, and the whole heated. Then the albumen will be precipitated and should be removed while hot by filtration, while the filtrate is allowed to cool off. If a cloudiness now arise on the addition of salt solution to the filtrate then hemialbumose is present. If too much salt solution be added, the precipitate of hemialbumose cannot be redissolved by heat.

Peptones are present in the urine principally in the absorption of pus and exudations (pneumonia, empyema, abscesses and puerperal fever, etc.) They are not precipitated on heating, nor with nitric nor acetic acid, nor with ferro-cyanide of potassium. They are tested for with the biuret test after the *albumen and hemialbumose have been removed or proved absent.*

10 ccm. [2½ drachms] of a concentrated solution of sodium acetate and a few drops of a solution of iron chloride are added to 500 ccm. [1 pint] of urine until there results a permanent red color ; then a caustic potash solution is dropped carefully into this mixture until it is slightly acid or neutral, and the mixture heated. After it has cooled off and been filtered, the filtrate, which ought to be entirely free from albumen, is subjected to the biuret test.

Blood.—We speak of hæmaturia when the blood coloring matter is present in the urine in combination with the blood corpuscles ; of hæmoglobinuria when the blood coloring matter is in solution without there being blood corpuscles in the sediment. The latter occurs when the blood corpuscles are dissolved by some agent (after poisoning, cold, etc.), and the hæmoglobin becomes free.

Urine containing blood-coloring matter is either bright red with a greenish iridescence (resembling meat juice) from the presence of oxyhæmoglobin, or it is a dark brownish-red from the presence of metahæmoglobin. The latter differs from oxyhæmoglobin by its being recognized in the spectroscope as a dark absorption line in the red and a paler one between the green and blue near both the oxyhæmoglobin lines.[1]

---

[1] The spectroscopic examination may be made with the pocket spectroscope. The urine is held in a tube before the slit in the instrument.

Besides, the spectroscopic test blood-coloring matter may also be recognized by the following tests :

Heller's Test.—If the urine be heated with caustic potash, the earthy phosphates in precipating take the coloring matter of the blood with them, and appear reddish brown instead of white.

Guaiac Test.—About 1 ccm [15 drops] of a freshly made tincture of guaiac and the same amount of resinous turpentine oil are added to some urine and well shaken. If blood be present the mixture turns blue after a few minutes. Instead of turpentine oil, Hühnerfeld's mixture [1] may be used.

The smallest amount of blood which can no longer be recognized by one of these methods may be looked for by *examining the sediment microscopically* for blood corpuscles.

Coloring matter of the bile. In urine there is present either the actual coloring matter of the bile (bilirubin) which is changed by oxydation into green (biliverdin), violet, red, and yellow (choletelin), or *hydro-bilirubin* (urobilin), which originates from a reduction of the coloring matter of the gall and blood. Urine containing bilirubin is of a beer-brown color and has a yellow foam on shaking. On being shaken with chloroform the bilirubin becomes gold-yellow and is taken up by the chloroform.

Bilirubin is tested for by the *Gmelin* test. A few drops of fuming nitric acid are added to concentrated nitric acid until a slight yellow is observed. This mixture is then poured into a vessel containing urine in such a way that the acid passes down the side of the glass under the urine. Then there is formed at the point of contact of the acid and the urine, a colored ring which passes from green through violet to red and yellow. A blue ring alone may be caused

---

[1] Glacial acetic acid, 2.0 ccm [30 drops].
Distilled water, 1.0 ccm [15 drops].
Oil of turpentine,
Absolute alcohol,
Chloroform,—of each 100.0 ccm [3 ounces].

by indigo, a reddish-brown one by hydrobilirubin and other substances.

If a solution of iodine in iodide of potash be added to the urine containing bilirubin, it becomes a green (biliverdin).

Hydrobilirubin is tested for by adding to urine 2–5 drops of a 10 % solution of the chloride of zinc, and afterwards enough ammonia to redissolve the precipitated oxide of zinc. If a green fluorescence is observed (by looking at the test-tube against a dark background) in the fluid filtered from the precipitated phosphates, hydrobilirubin is present. Instead of the chloride of zinc and ammonia, iodine-iodide of potash and caustic potash may be used. In the spectroscopic examination of urine containing hydrobilirubin (even after adding the chloride of zinc and ammonia), it may be recognized by an absorption line between the green and the blue.

Gallic acids are found by *Pettenkoffer's test:* A grain of cane sugar is added to the fluid and the whole is evaporated with gentle heat on the cover of a porcelain crucible with a drop of concentrated sulphuric acid. If the gallic acids are present the fluid becomes purple. The same reaction may be caused by other substances (albumen, fatty acids, etc.), so that the gallic acids should first be extracted from the urine. For the procedure necessary (as evaporating, extracting with alcohol, precipitating with baryta and extracting the cholalate of baryta with warm water), see the text-books.

Grape Sugar (Dextrose) $C_6H_{12}O_6$ is fermented by yeast to alcohol and carbonic acid ($= 2C_2H_5OH +$ $2CO_2$), shows a brown color when heated with caustic potash, is capable of reducing, and turns the plane of polarized light to the right.

I. To make the fermentation test, a test-tube or eudiometer tube is first half filled with mercury, and the same amount of urine is added, only leaving enough room for a little yeast. The air bubbles are removed from the opening of the tube, which is then closed by the finger, and dipped upside down under mercury, and left there at a temperature not over 30° C [86° F], The presence of grape sugar soon causes a development of gas. In order to show that the gas is carbonic acid, some caustic potash is introduced through a curved pipette into the tube, and by this the carbonic acid is absorbed.

This determines the presence of grape sugar. Much more convenient are the so-called *fermentation tubes*.[1] A piece of yeast as large as a pea is introduced into one of the tubes and urine is so added that no air enters into the vertical branch of the tube. For the sake of greater certainty a second tube with a dextrose solution and yeast, and a third tube with normal urine and yeast, may be also used. If the result of the second test is positive, this shows that the yeast is effective, and if the result of the third test is negative, it shows that the urine contains no sugar.

By determining the specific gravity of urine both before and after fermentation (after 24 hours at the temperature of the room), the approximate amount of grape sugar may be obtained. The urine is made to ferment with yeast in a long-neck bottle, the opening being covered with a watch glass to prevent evaporation. After 24 hours the specific gravity of the filtered urine is taken at the same temperature. The difference in the specific gravity before and after the fermentation is read from the urinometer, each degree of which corresponds to 0.219 % of sugar. Thus urine which before the fermentation had a specific gravity of 1040, and after, the specific gravity of 1020, contains 4.38 % of sugar.

II. *Moore's Test.* If urine containing sugar be heated a *few minutes* with one third its volume of a concentrated caustic potash solution, it turns brown. This test is reliable only when the brown color is very intense, for sugar to the amount of 0.5 % cannot thus be found. With 1 % of sugar the color becomes canary yellow, 2 % amber yellow, 5 % the color of Jamaica rum, and 7 % it becomes blackish brown and non-transparent.

III. *Reduction Tests.*

(a) *Trommer's Test.*—To a quantity of urine, one third its volume of a caustic potash or soda solution is added, and then 1–2 drops of a diluted (5–10 %) sulphate of copper solution. If the bright blue-colored precipitate of hydrated copper oxide remains undissolved and flocculent on shaking, no sugar is present. In the presence of sugar, glycerine, tartaric acid, ammonia, or albumen the hydrated cupric oxide dissolves, giving the urine a sky-blue color. The sulphate of copper solution should be added drop by drop until there

---

[1] To be had of Hildenbrand in Erlangen, and Dr. R. Muenke in Berlin, N. W., Luisenstrasse, 58.

is only a small part left undissolved on shaking the tube. If this mixture be then heated the presence of sugar will cause, before the boiling point is reached, a yellow-red precipitate of cuprous oxide (Cu_2O), formed by the grape sugar taking oxygen from the cupric oxide (CuO). If the fluid change color without forming a precipitate, or if the latter be not formed until the urine has cooled off, then the test is not convincing, since other reducing substances (uric acid, creatinin, etc.) hold the cuprous oxide in solution. Exceptionally reducing substances appear in the urine from medicines taken (turpentine, chloral hydrate, chloroform, benzoic acid, salicylic acid, camphor, copaiva, and cubebs). It is generally a more certain test to let the urine stand for 24 hours, cold, after adding the substances, instead of heating it. If then a yellow precipitate of cuprous oxide appear, it can be caused by sugar alone.

(*b*) *Test with Fehling's solution.*—Fehling's solution consists of

Crystalline sulphate of copper 34.639 [520 grains].
Neutral tartrate of potash 173.0 [5½ ounces].
Officinal caustic soda solution 100.0 [3 ounces].
Distilled water enough to make 1000.0 [30 ounces].

One ccm. [15 drops] of this is exactly reduced by 0.005 gram [$\frac{1}{12}$ grain] of grape sugar. Two ccm. [30 drops] of this fluid are put into a test tube, diluted with an equal amount of water and heated. In case the formation of the oxides takes place, which would make it unfit for use, a few ccm of urine which have been previously heated in another test-tube are added to this. If grape sugar be present, a yellowish-red precipitate is formed.

In order to *approximately determine quantitatively the amount of grape sugar* present, the trituration method of Fehling may be carried out on a small scale. Two ccm. [30 drops] of Fehling's solution (corresponding to 0.01 gram [$\frac{1}{6}$ grain] of sugar) are measured off in a large test-tube and diluted with about ten times its volume of water and heated. By means of a dropper 1–3 drops of urine are then added, and the whole is heated, observing whether the fluid still shows a blue color on holding it to the light. Is this the case, then a few more drops are added, and it is again heated and again observed and then watched until the last trace of blue has just completely disappeared, showing that all the cuprous oxide has been reduced. We know that in the urine there is exactly 0.01 gram [$\frac{1}{6}$ grain] of sugar, and counting 20 drops to 1 ccm. we can calculate the percentage of sugar present. In order to save the time and trouble of making such calculation at every examination, the following table will be found convenient. It is better to dilute the urine four or five times in a graduated glass.

Drops = % Sugar.		Drops = % Sugar.		Drops = % Sugar.	
1	20	10	2.0	25	0.8
2	10	11	1.8	30	0.6
3	6.6	12	1.6	40	0.5
4	5	13	1.5	50	0.4
5	4	14	1.4	60	0.3
6	3.3	15	1.3	70	0.28
7	2.8	16	1.2	80	0.25
8	2.5	18	1.1	90	0.21
9	2.2	20	1.0	100	0.20

(c) *Böttger's Test.*—The urine is made alkaline by saturating it with sodium carbonate in substance, adding a pinch of the subnitrate of bismuth (NO_3BiOH_2) and heating it a few minutes. Or the urine may be heated with $\frac{1}{10}$ of its volume of Nylander's solution. This solution consists of neutral tartrate of potash 4.0 grams [1 drachm], 10% solution of caustic soda 100 ccm. [3 ounces], to which are added, subnitrate of bismuth 2.0 grams [30 grains] while warm, and the whole to be filtered after cooling off. In the presence of grape sugar a brown or black color is formed, due to the metallic bismuth.

(d) *Mulder's Test.*—The urine is first made alkaline with carbonate of sodium, and then a solution of indigo carmine (sulphate of indigo) is added until the urine turns blue. On heating, the indigo blue is reduced by the grape sugar present to indigo white, and on exposure to the air again, turns blue.

IV. *Test with phenylhydrazin.*—Two pinches of phenylhydrazin and four pinches of the acetate of sodium are put into a test-tube, which is then half filled with water and heated. Then an equal volume of urine is added, and the test-tube is heated for 20 minutes in a water bath and allowed to cool off. When the urine contains a large amount of grape sugar, a yellow crystalline precipitate of phenylglucosazone is formed, and when there is only a little grape suger present the sediment under the microscope shows these crystals of this form (v. Jaksch).

V. *Polariaztion Test.*—The specific angle of grape sugar for yellow sodium light (α) D is 53°. From the degree of deviation α in the special case, and the length l of the tube used expressed in decimeters, the percentage of grape sugar in the urine may be calculated from the formula $p = \dfrac{\alpha.100}{53.l}$

With the presence of substances turning the light to the left, as albumen or oxybutyric acid, the determination by polorization is of little value ; therefore it is best to ferment the urine and then polarize it a second time. Dark or cloudy urine should be made clear by the addition of $\frac{1}{10}$ its volume of sugar-of-lead solution in a measure glass, and the dilution should of course be taken into account.

Sugar of milk (lactose), $C_{12}H_{22}O_{11}$, is present in the urine of nursing lying-in women. It has a right rotatory power $(\mu)_D = 52.5$, and passes over with difficulty into alcoholic fermentation and rarely into lactic acid fermentation. It has the property of reduction.

Inosite, $C_6H_{12}O_6$, is present in polyuria. It is neither fermentable, nor does it possess the power of polarization nor of reduction. For its formation see the textbooks.

Acetone, CH_3COCH_3, is present in urine in febrile diseases, in diabetes, in certain forms of carcinoma, and in inanition and auto-intoxication.

To test for the acetones a few drops of freshly-prepared nitroferrocyanide of sodium are added to the urine, and then a strong caustic soda solution, until it is decidedly alkaline. When the beginning purple tint turns yellow, 1 to 3 drops of concentrated acetic acid are added, and if the acetones be present a crimson-purple color is formed at the point of contact of the acetic acid and the mixture (Legal's test). It is better to distill the urine with some muriatic acid, and to test the distillate for acetone with Lieben's test. According to the latter a few drops of a solution of iodine in iodide of potash and caustic potash are added to a **few** ccm. of the distillate. If the acetone be present, a yellow-white precipitate of iodoform appears at once.

Diacetic acid, CH_3COCH_2COOH, is present in the urine in many grave cases of the contagious diseases, in grave cases of diabetes, carcinoma, and in auto-intoxication.

If to some urine one or two drops of a solution of the chloride of iron be added, a gray or chocolate-colored precipitate of the phosphate of iron appears. If more iron chloride be added, the presence of diacetic acid gives the urine a *dark Bordeaux-red color* (the iron chloride reaction of Gerhardt), which disappears at once on adding sulphuric acid. If the urine be heated first, then the reaction is very slight or not at all. If the urine already made acid with sulphuric acid be extracted with ether, the ether takes up the diacetic acid, and may be tested for with iron chloride. Still even this reaction disappears in 24–48 hours. A brown-red color of the urine with iron chloride does not determine the presence of diacetic acid. If the urine be distilled the diacetic acid splits up into acetone and carbonic acid, and the acetone may then be determined by Lieben's test.

Diazoreaction (Ehrlich). Sulphodiazobenzole unites with different kinds of unknown aromatic substances of the urine to form colored compounds.

To prepare this reagent two solutions are necessary :

a) Sulphanile acid . . . .   5.0 [75 drops].
   Muriatic acid . . . .   50.0 [1½ ounces].
   Distilled water . . . . 1000.0 [30 ounces].

And

b) Nitrite of sodium . . . .   0.5 [8 grains].
   Water . . . . 100.0 [3 ounces].

When ready for use 5 ccm. [75 drops] of solution *b*) are to be added to 250 ccm. [8 ounces] of solution *a*), and this " reagent " should be prepared fresh for every test. Then to equal parts of this reagent and urine ⅛ volume of ammonia is added, and shaken up. In certain (febrile) diseases the fluid turns red (scarlet, orange, orange-red), which is especially noticeable in the foam (red reaction). This color is noticeable in typhoid fever (from the first week on), sometimes in relapses, also in grave cases of phthisis pulmonum, pneumonia, measles. The disappearance of this reaction is considered a good sign.

Melanine.—In the urine of those suffering from melanotic carcinoma, melanogen is sometimes present, which forms black clouds of melanine on adding concentrated nitric acid or chromic acid to the urine. At times the urine is dark from the presence of formed melanine in the urine.

Sulphuretted hydrogen, H_2S, is present principally in decomposed urine, as in cystitis. Since it is found in normal urine after long standing, only fresh urine should be taken. A few drops of muriatic acid are added to the urine in a bottle, and the opening is covered with filter-paper which has been moistened with a sugar-of-lead solution. If H_2S be present the moistened paper turns dark from the formation of the sulphide of lead.

Leucine or *Amidocaproic acid*, and *Tyrosine* or *Amido-Hydroparacumaric acid* are present in the urine principally in acute yellow atrophy of the liver and in phosphorus poisoning. Leucine appears in yellow globules, which have a fatty gloss and are often marked with radiating lines. Tyrosine is in the form of fine bundles of

Fig. 30.   Fig. 31.

Leucine.   Tyrosine.

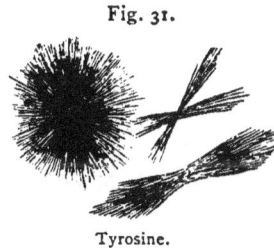

needles or globules. The urine is evaporated to syrupy consistency and left in the cold to crystallize, and then examined microscopically.

Cystine is occasionally present in the sediment in the form of colorless shining hexagonal plates.

Fat is occasionally present as a fine cloudiness and gives the urine a milky appearance (chyluria). This milkiness disappears on adding caustic potash and shaking up with ether.

Test for Drugs.

Iodine and *Bromine.*—Freshly made chlorine water or strong fuming nitric acid is added to the urine and then shaken with a few ccm. of chloroform, which is then colored carmine red if iodine is present, and brownish yellow if bromine is present.

Nitric acid.—A brucine solution is added to the urine and the sulphuric acid is allowed to trickle down the side of the glass and at the point of contact a red ring is formed. The same reaction may be caused by other bodies (as hydrobilirubin).

Lithium.—The flame reaction or a spectroscopic examination of the ash is sufficient.

Arsenic.—After removing the organic substances with muriatic acid and chlorate of potash, the fluid is examined according to Marsh's test.

Lead.—Fresh muriatic acid and chlorate of potash are added to destroy the organized substances, the chlorine is driven off, then the mixture is filtered off and sulphuretted hydrogen conducted through it, and if lead be present a brown color, due to the sulphide of lead, is formed.

Mercury.—To the urine of one day, 10 ccm. [2 drachms] of muriatic acid and a small quantity of brass or copper shavings are added, and the whole is heated. After 24 hours the urine is poured off and the metal washed several times in water made slightly alkaline with caustic potash, then washed with alcohol, then with ether, and then let dry. The metal is then brought into a long large dry test-tube and heated red-hot. If the mercury be present it has already amalgamated itself with the copper or brass and the heat volatilized it and caused it to be condensed on the cool parts of the tube. Now, if fumes of iodine be introduced into the tube the mercury is changed to the iodide of mercury which appears as a red tinge, and by careful heating may be condensed to a sharply defined ring.

Quinine.—500 cm. [15 ounces] are made alkaline with caustic potash and shaken five minutes with ether. The ether is then brought to the surface and evaporated off, and the remainder taken up with water and a few drops of muriatic acid. This fluid shows a blue fluorescence on adding a drop of sulphuric acid, or, if treated with strong chlorine water and concentrated ammonia, it shows a green ring.

Carbolic acid (Phenol C_6H_5OH) — When much carbolic acid has been ingested the urine becomes greenish brown and turns dark when exposed to the air, just as the urine does after taking hydroquinone $(C_6H_4(OH)_2)$, folia uvæ ursi, and tar. For the behavior of sulphuric acid in carbolic acid intoxication, and the detection of carbolic acid, see pages 73 and 75.

Salicylic acid (Oxybenzoic acid).—The urine turns violet on adding chloride of iron.

Antipyrin.—The urine turns red on adding chloride of iron.

Thallin.—The urine is greenish-brown and turns purple on adding iro⁻ chloride. On shaking up the urine with ether, the unchanged thallin is also taken up by it and this turns green on adding the chloride of iron.

Kairin.—The urine is greenish brown, turns dark on standing, and turns brownish-red on the addition of the chloride of iron.

Turpentine.—The urine smells of violets and a precipitate is sometimes formed on adding nitric acid.

Tannin.—The urine turns bluish-black on the addition of chloride of iron.

Santonin.—The urine is straw-yellow and turns scarlet on the addition of alkalies.

Rhubarb and *Senna* (Chrysophanic acid).—The urine turns also red on the addition of an alkali, but the color remains permanent, while in the case of santonin it soon disappears. On the addition of baryta water the precipitate with rhubarb and senna is red, and with santonin the filtrate. Ether takes up the color of senna and rhubarb, but not of santonin.

Organic Sediments.

Leucocytes are present normally in a small number in the urine, and in a large amount in inflammation and suppuration in any part of the genito-urinary apparatus (nephritis, pyelitis, cystitis, gonorrhœa, fluor albus). In alkaline urine the pus is of a mucous nature. *Red blood corpuscles* in the urine are generally free from color, and in renal hemorrhage are often in the casts.

The *renal epithelium* is small, round, or cuboid with a vesicular nucleus, and often very full of fat drops. They are often arranged in cylindrical form or lie on the tube casts adherent to them (epithelial casts). The appearance of renal epithelium in the urine always points to a

Fig. 32.

morbid process in the kidney. When there are numerous fatty degenerated epithelial cells in the urine, it is a sign of chronic parenchymatous nephritis.

Renal Epithelium partly undergoing fatty degeneration.

The *epithelial cells of the bladder, ureters, and renal pelvis* do not differ from each other in appearance. The cells of the superficial layers have a polygonal form, those of the deeper layer are somewhat round, often with processes (pear-shaped), and contain a vesicular nucleus. If there are many of such cells with leucocytes in the urine, it is evidence of an inflammatory condition of the bladder, ureters, or renal pelvis. The

Fig. 33.

Superficial layer.

Deep layer.

Epithelium of the bladder, urethra, and renal pelvis.

microscopical examination is of no assistance here, but we can generally take it for granted that the urine in pyelitis is generally acid, and in cystitis generally alkaline.

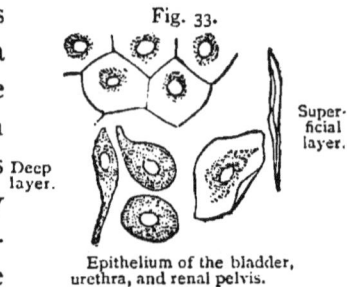

The *vagina* and *prepuce* possess very long, flat epithelial cells like those of the mucous membrane of the mouth. The male urethra has cylindrical epithelium. These epithelial cells are often found in the suppuration of acute gonorrhœa. The gonorrhœal pus is also characterized by the presence of gonococci (see Chapter XI.).

Casts are effusions into the urinary tubules. They are present in all cases of albuminuria, not only in nephritis but also in all irritative conditions of the kidneys (icterus, the acute contagious diseases, heart diseases, etc). We distinguish (1) *Hyaline casts*, which consist of a homogeneous translucent substance, and possess a very delicate contour, which is often scarcely visible. (2) *Granular casts*, having a fine-grained substance, but otherwise resembling the hyaline casts. (3) *Waxy casts*, of a yellow color and greater lustre, with sharply-defined contour, and are often irregularly curved and bent. They are found principally in chronic nephritis, and point to a grave disturbance. (4) *Brown casts* are present in fractures, and also in the grave cases of contagious diseases. (5) *Cylindrical casts* are long, irregularly broad, with long stripes on them. These are perhaps only mucous threads, and are of no diagnostic importance. Very often other substances are attached to the casts, especially to the hyaline casts, as urates, fat drops, red blood corpuscles, leucocytes, and renal epithelium.

Further, there are found occasionally in the urine spermatozoa and cells of neoplasms (cancer, papilloma).

Micro-organisms are present in fresh urine in several of the contagious diseases (diphtheria, recurrent fever), in cystitis, and pyelonephritis (in a cylindrical form), also tubercle bacilli in tuberculosis of the genito-urinary tract and gonococci in gonorrhœa.

Animal Parasites :

(1) Echinococcus cysts and hooklets.

(2) Embryos of filaria sanguinis : small snake-like worms which are exceedingly movable, and are as broad as the diameter of a red blood corpuscle, and 0.35 mm. [$\frac{1}{150}$ inch] long.

(3) Distomum hæmatobium whose eggs have on one end or on the side a spinous process. The two last parasites may cause hæmaturia and chyluria (*v.* Chapter XI.).

For the analysis of the urinary concrements see Chapter XIII.

CHAPTER X.

TRANSUDATIONS AND EXUDATIONS.

THE different *serous transudations* have a very different specific gravity according to their origin. They are in the order of their specific gravity ; hydrocele, hydrothorax, ascites, anasarca, and hydrocephalus.

The serous (inflammatory) *exudations* have a greater specific gravity than the simple transudations of congestion, and indeed it may generally be taken for granted that a fluid, be its origin what it may, is the product of an inflammation when its specific gravity exceeds 1018 (pleurisy, peritonitis), and that it is simply a transudation due to congestion when its specific gravity

in hydrothorax	is	less	than	1015,
" ascites	"	"	"	1012,
" anasarca	"	"	"	1010,
" hydrocephalus	"	"	"	1008.5.

Now since the amount of ash, extractive matter, etc., contained in exudations and transudations varies very slightly and the amount of albumen varies very greatly, we conclude that the specific gravity is principally dependent upon the amount of albumen contained in these fluids. Therefore from the specific gravity the amount of albumen may be approximately determined according to the formula of Reuss,

$$E = \tfrac{3}{8} (S - 1000) - 2.8.$$

in which E denotes the amount per cent. of albumen sought and S the specific gravity. Accordingly in a specific gravity of 1018, 3.95 % of albumen would be calculated. These rules hold good for serous exudations, but not for purulent, chylous, and very hemorrhagic exudations, nor for those in diabetes, cholæmia, and uræmia.

In order to determine the *specific gravity*, the fluid should be protected from evaporation and cooled off to the surrounding temperature, since fluid at body temperature has too low a specific gravity; for every 3° Celsius [5.4° Fahrenheit] increase corresponds to about one degree of the aræometer less.

The *amount of albumen* is determined by diluting a known quantity of the exudation (10 ccm. [2½ drachms])[1] with ten times its volume of water, heating it to the boiling point and adding dilute acetic acid drop by drop until the fluid is slightly acid. The precipitate of albumen is then to be collected upon a filter paper which has previously been dried at a temperature of 100° C. [212° F.] and weighed, washed with water, then with alcohol and ether, the total weight to be deducted from the weight of the filter paper. The filtrate should be clear and free from albumen, which may be proved by adding a few drops of ferrocyanide of potash to the liquid.

Exudations and transudations have an alkaline reaction, and deposit, on standing, a more or less abundant amount of fibrin. A microscopical examination reveals in the coagulum leucocytes and swollen endothelial cells, which often contain vacuoles.

The contents of the *echinococcus cysts* are generally clear, neutral, or alkaline, and the fluid has a specific gravity

---

[1] To determine the amount of albumen in urine 50 or 100 ccm. [1½–3 ounces] of urine should be taken.

of 1008–1013, contains little or no albumen, but chloride of sodium in large quantities, as well as grape sugar and succinic acid. The latter is detected by evaporating the fluid, acidifying it with hydrochloric acid and shaking it up with ether, and after the evaporation of the ether the succinic acid remains as a crystalline mass, whose water solution with the chloride of iron forms a gelatinous, rust-colored precipitate of succinate of iron. When heated in a test-tube the irritating fumes of the succinic acid are given off, causing cough.

On microscopic examination the scolices and ring of hooklets are sometimes found. In the older lifeless cysts are found crystals of cholesterine and hæmatoidine.

The water of *hydronephrosis* is generally clear, of a specific gravity of 1010–1020, contains mucus, sometimes blood and pus, and a varying amount of albumen and of urinary constituents. But since these are also found in the fluid of the echinococcus the diagnosis of hydro-nephrosis should be made only when there is a larger amount of urea and uric acid present. Urea is detected according to the method on page 70 ; uric acid, by adding muriatic acid and examining microscopically the crystals formed, or by the murexide test.

The pear-shaped epithelial cells of the renal pelvis and tube casts are also occasionally present.

The contents of an *ovarian cyst* are generally mucous, tenacious, yellow ; but may be watery, semi-fluid, and brown. The specific gravity is between 1003 and 1055 and generally between 1010 and 1024. The fluid usually contains albumen and metalbumen (pseudo-albumen), which causes the mucous consistency. This is not pre-cipitated by acetic acid (differing in this respect from mucin), nor by heat, nor by nitric acid ; but falls into fibrous flakes on adding alcohol. By heating it with the mineral acids, a reducing substance is formed.

To detect the metalbumen the fluid is freed from albumen by heat and acetic acid. When metalbumen is present the filtrate is opalescent and mucous. It is precipitated into white flakes on adding alcohol in excess. The flakes are then pressed out and heated .with dilute muriatic acid (5 %) until they turn brown ; after cooling off they are made alkaline with caustic soda, and a few drops of a cupric sulphate solution are added and the whole heated. If metalbumen be present there is a precipitate of yellow cuprous oxide.

A microscopic examination occasionally shows the presence of cylindrical and ciliated epithelium, and sometimes colloid particles.

CHAPTER XI.

PARASITES.

Animal Parasites.

Cestodes.—*The tape-worms* represent colonies of individuals which consist of a head with hooklets and of a larger or smaller number of single individuals called proglottides or segments. The eggs which come from the matured proglottides (hermaphrodite), if they come into the stomach of the right animal, develop in its organs into a cysticercus. If this cysticercus be taken into the intestinal canal, it becomes a tapeworm.

Fig. 34.[1]	Fig. 35.	Fig. 36.	Fig. 37.	Fig. 38.
Segment of Tænia solium.	Segment of Tænia saginata.	Segments of Bothriocephalus latus.	Egg of Tænia solium.	Egg of Bothriocephalus latus.

Tænia Solium is 1–3 metres [yards] long. The head is as large as a pin's head, has four suckers, a rostellum or proboscis upon which there is a double row of hooks.

---

[1] Figs. 34, 35 and 36 are from Stein's Entwickelungsgeschichte und Parasitismus der menschlichen Cestoden.

The matured proglottides have the sexual openings on the side and a uterus with 7 to 10 thick lateral branches, which subdivide (fig. 34). The eggs are round or oval, with a striped shell and an embryo having six hooks (fig. 37). The *cysticercus cellulosæ* is about as large as a pea, and is found in swine and in man (when the eggs are taken into the stomach) under the skin, in the muscles, in the brain, eye, etc.

Tænia saginata or mediocanellata is thicker and larger than the former. It has a head with four suckers, but no rostellum and no hooks. The proglottides have lateral sexual organs and a uterus which subdivides into 17 to 30 finer branches (fig. 35). The eggs are like those of the tænia solium, only somewhat larger. The cysticercus is smaller, and is found in the flesh of cattle (also in deer and sheep).

The Bothriocephalus latus is 5–9 metres [yards] long, and has a lancet-shaped head with two lateral grooves. The matured segments are broader than they are long. The uterus has a brownish tinge, and is arranged in the form of a rosette around the flat sexual openings (fig. 36). The eggs are oval and have a cover. The cysticerci are found in fish (salmon).

Tænia nana, tænia flavopunctata and tænia cucumerina (elliptica) occur sporadically in man.

Tænia echinococcus is found in the dog. It is 2½ to 4 mm. [⅛–1/16 of an inch] long, has a head with hooklets and suckers, and three segments, of which the last one only is matured. The cystic form of the echinococcus is found in man (in the liver, spleen, kidneys, lungs, etc.). It is observed in two forms, as a large echinococcus sac filled with daughter cysts, and as an echinococcus multilocularis, which consists of a very large number of minute cavities filled with a gelatinous substance and with concentrically arranged walls. In the echinococcus cysts, heads (scolices) with hooks are sometimes found. (For the echinococcus fluid see page 93.)

Nematodes or **Round Worms** are bisexual.

The *Ascaris lumbricoides* or round worm has its habitat in the small intestine. It resembles the rain worm. The male is somewhat smaller (150–250 mm. [4–6 inches]) than the female (150–250 mm. [4–6 inches]), and its head is rolled up. The eggs, which are evacuated in large numbers with the stools, have a thick, concentric-

Fig. 39.	Fig. 40.	Fig. 41.	Fig. 42.

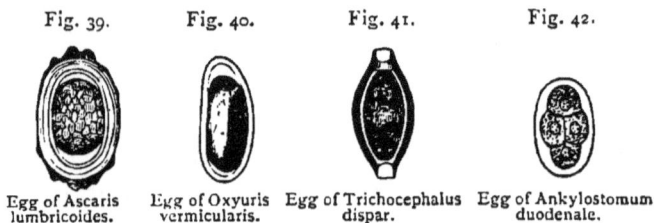

Egg of Ascaris lumbricoides.	Egg of Oxyuris vermicularis.	Egg of Trichocephalus dispar.	Egg of Ankylostomum duodenale.

ally striped shell, upon which lies a projecting albuminous cover (fig. 39).

The Oxyuris vermicularis or small thread-worm is found in both large and small intestines. It often passes from the intestine to the anus, causing violent itching in that region. The male is 3–5 mm. [⅛–¼ inch] and the female 10 mm. [½ inch] long, the former having blunt ends, and the latter being pointed. The eggs, which are especially numerous around the anus of man, are oval and possess a thin shell (fig. 40).

The *Trichocephalus dispar* or whip-worm lives in the large intestine ; it is 4–5 cm. [1½–2 inches] long, has a thread-like head extremity, and a thicker spirally rolled body in the male, and a straight slightly curved body in the female. The eggs are yellow in color and shaped like a lemon (fig. 41).

The *Anguillula intestinalis* (Rhabdonema strongyloides, Leuckart) is 2.2 mm. [$\frac{1}{8}$ inch] long and lives in the upper part of the small intestine. The eggs, which resemble those of the ankylostomum duodenale, grow in the intestine to larvæ 0.2 mm. [$\frac{1}{80}$ inch] long which are found in the fæces as small worms with lively movements. Outside of the body the latter are developed to an intermediate form, the anguillula stercoralis. This belongs to the developmental cyclus of the anguillula intestinalis.

The *Ankylostomum duodenale* lives in the small intestine, and causes anæmia by boring into the intestinal wall (*e. g.*, in the anæmia of the St. Gotthard tunnel workmen, brickmakers, and miners). The male is 10 mm. [$\frac{1}{2}$ inch] long and the female 12–18 mm. [$\frac{1}{2}$–1 inch] long. The eggs, which are passed in large numbers with the stools, have clear, simply formed shells and an embryo which is generally undergoing fission (fig. 42). The eggs are developed a few days only, after the passage of the larvæ from the intestines.

The *Trichina spiralis* enters the intestine through trichinosed pork. The male is 1.5 mm. [$\frac{1}{13}$ inch] long and the female 3 mm. [$\frac{1}{7}$ inch] long. The matured worms live in the small intestine and bring forth, after 5–7 days, young trichinæ, which then bore through the intestinal walls, get into the circulation, and fix themselves in the course of the next few days in the muscular fibres, where they may become encapsuled. Their presence at first causes fever.

The *Filaria sanguinis* is found in the tropical regions. It causes hæmaturia, chyluria, and disturbances of the lymph circulation. The matured form lives in the lymphatic organs of man, and here gives rise to a large number of living embryos, which are found in the urinary sediment and blood, and indeed in the latter in such large quantities that every drop of blood contains several embryos. These appear as little worms which move freely, and are surrounded by a delicate envelop. They are 0.35 mm. [$\frac{1}{130}$ inch] long and as broad as the diameter of a red blood corpuscle.

The *Filaria medinensis* may reach 80 cm. [$30\frac{1}{2}$ inches] in length and $\frac{1}{2}$–$1\frac{1}{4}$ mm. [$\frac{1}{30}$–$\frac{1}{60}$ inch] in breadth. It occurs in the tropics, and leads to the formation of abscesses of the skin.

Trematodes or Flat Worms.

The *Distomum hepaticum* is 28–32 mm. [1 inch] long, of a leaf-like form, with conical-shaped forepart of the body. The eggs are very large 0.13 mm. [$\frac{1}{20}$ inch] long (see fig. 43), and with a cover.

Fig. 43.   Fig. 44.

Egg of
Distomum
hepaticum.

Egg of
Distomum
hæmatobium.

The *Distomum lanceolatum* is smaller than the former. It may reach 9 mm. [$\frac{3}{8}$ inch] in length. The eggs are, likewise, considerably smaller. Both are found in the gall bladder and gall ducts. The eggs are sometimes found in the fæces.

The *Distomum hæmatobium* occurs in the tropics. It lives in the abdominal veins, and causes diarrhœa, hæmaturia, and chyluria. The male is 12–14 mm. [$\frac{1}{2}$ inch] and carries in a groove in it the female, which is 16–19 mm. [$\frac{5}{8}$–$\frac{3}{4}$ inch] long. The eggs are 0.12 mm. [$\frac{1}{20}$ inch] long, are found in the urinary sediment, and have a point either at one pole or on the side (fig. 44).

Arthropodes.

Acarus (sarcoptes) *scabiei*, or itch-insect is an oval lenticular body with eight short legs. The female is found at the end of the furrow, which is filled with the eggs and excreta of the insect. In 8 to 14 days the young ones are hatched, and in turn bore into the skin. *Acarus* (demodex) *folliculorum* is longer than it is broad, and is found in comedones, especially in the face. *Pediculus capitis*, or head louse ; *Pediculus vestimenti*, or body louse ; *Pediculus pubis*, or crab louse. *Pulex irritans*, or flea.

Protozoa.

In the stools the following protozoa are sometimes found in chronic diarrhœa :

Amœba coli, a round granular structure with a nucleus and a few vacuoles. *Cercomonas intestinalis*, pear-shaped animalculæ (8–10 μ [$\frac{1}{30}$–$\frac{1}{25}$ inch] long), with ciliated extremities. *Trichomonas intestinalis* (10–15 μ [$\frac{1}{25}$–$\frac{1}{10}$ inch] long), almond-shaped with ciliated extremities. *Balantidium* or *Paramœcium coli*, pear-shaped, 70–100 μ [$\frac{1}{4}$–$\frac{1}{3}$ inch] long, ciliated, with an inverted mouth. Besides these protozoa, there are found in the vaginal secretion, trichomonas vaginalis, and in other secretions other protozoa.

Vegetable Parasites.

Hyphomycetes.

Achorion Schoenleinii, or *favus fungus*, is in the shape of worm-like filaments, which are provided with septa and lateral elevations, and in their ends are round or oval, brightly shining spores (conidia).

The *Trychophyton tonsurans* is the fungus of *herpes tonsurans* and circinatus, as well as of acne mentagra (sycosis parasitaria). The mycelium consists of curved and branching filaments provided with septa. The filaments have partly at their ends shining spores (conidia) with double contour. In the epidermis the fungus filaments are found, while in the hair and hair-sheath the spores (conidia) are found.

The *Microsporon furfur*, or fungus of *pityriasis ver*‧

sicolor, is found in the yellowish epidermis scales in large numbers as a dense network of curved, more or less branched filaments, with heaps of shining spores (conidia) within.

The *Microsporon minutissimum* is a very fine non-branching filamentous fungus without the formation of spores, and is found in erythrasma, but whether in causal relation or not is doubtful.

The *oidium albicans*, or thrush fungus, is found in the mouth cavity as well as in the œsophagus and the stomach. It consists of branching filaments, with shining, round or oval spores (conidia) at the points of bifurcation.

The *Aspergillus glaucus* and *niger* are often found in the sputum of consumptives or imbeciles, and may cause a peculiar kind of pneumonia called pneumonomycosis aspergillina They are filaments more or less branched, with double contour and with many brown pigmented spores.

In order to make the filamentous fungus visible, the preparation (from a scraped tongue, epidermis scales, hair, etc.) is allowed to stand for a few minutes in a 10 % caustic potash solution, which makes the albuminous substances and epidermis more translucent, and the fungus thereby all the more distinct.

Yeast fungi are frequently found in fermenting conditions of the stomach.

Schizomycetes or *Bacteria*.

Morphologically we distingnish

(*a*) The *Coccus* (spherical or oval), and according as the cocci are single, in twos, in chains, or in a racemose conglomeration, they are called monococcus, diplococcus, streptococcus, and staphylococcus.

(*b*) *Bacillus* or *rod*.

(*c*) *Vibrio*, or *curved rod*, fragments and developing form of spirilla as comma bacilli.

(*d*) *Leptothrix forms.* Filiform.
(*e*) *Spirillum.* Spiral form.

To the *Cocci* belong also the micro-organisms of erysipelas (round streptococci) and of puerperal fever, of gonorrhœa (bean-shaped diplococci which are found in clumps, partly in the leucocytes of the gonorrhœal pus), of croupous pneumonia and the pus-producing cocci, staphylococcus pyogenes aureus, the coccus of acute osteomyelitis, and staphylococcus pyogenes albus, etc.

To the *Bacilli* belong the micro-organisms of tuberculosis, of lepra (syphilis), of anthrax, malleus (glanders), typhoid fever, diphtheria, malignant œdema. As comma bacilli, are to be mentioned the cholera bacillus and the bacillus of Finkler-Prior.

To the *Spirilla* belong the Spirochæta Obermeieri, in recurrent fever, and the Spirochæta buccalis and others.

Figs. 45-54.

Fig. 45.
Bacillus mallei.

Fig. 46.
Bacterium pneumoniæ
crouposæ
(Friedlaender).

Fig. 47.
Bacillus
typhosus
(Eberth).

Fig. 48.
Bacillus
lepræ
(Hansen).

Fig. 49.
Bacillus tuberculosis
(Koch).

Fig. 50.
Bacillus anthracis.

Fig. 51.
Spirillum or spirochæta
Obermeieri
(recurrent fever).

Fig. 52.
Bacillus (s. spirillum) choleræ
asiaticæ (Koch).

Fig. 53.
Gonococcus
Neisseri.

Fig. 54.
Streptococcus
erysipelatis.

Clinically, the coloring of the micro-organisms in a *dried preparation is almost exclusively used.*[1]

[1] The preparation and the staining of the sections, as well as the methods of bacteria culture, are subjects too extensive to be taken up here. They may be better studied in the " Proceedings of the Royal Board of Health " (Berlin) ; Cornil et Babes : " Les Bactéries " ; Friedlaender : " Microscopical Technique " ; Hueppe : " The Methods of Bacterial Investigation."

A small drop or particle of the substance (blood, pus, sputum, tissue juice, etc.) to be examined is spread with a platinum needle upon a clean cover-glass, or two glasses are rubbed together so that a thin film of the matter is deposited on each. The cover-glasses are then to be protected from the dust and left until dry. Then the glass, with the preparation side turned upward, is passed three times moderately quickly through the spirit flame. Dry preparations of blood should be heated a few hours at a temperature above 100° C. (212° F.), in order to fix the hæmoglobin, and the best way to accomplish this, according to Ehrlich, is to put the glass on a metal plate, to one corner of which the heat is applied. The dried preparation may then be colored.

In clinical examinations aniline colors are principally used, and among them the following :

(*a*) The acid aniline colors : Eosine, picric acid (principally in blood examination).

(*b*) The basic aniline colors : Fuchsin (muriate of rosaniline), methyl blue, methyl violet, and gentian violet, vesuvin (Bismarck brown), and malachite green.[1] Of these colors it is well to have on hand either a concentrated, watery filtered solution, or, what is better in the case of fuchsin, a concentrated alcoholic solution.

The coloring of the dried preparations is carried out either by dropping with a glass rod some of the concentrated watery solution on the preparation, or, if the object is to let the color work in for a longer time, by letting the cover-glass float, with the preparation downward, on the surface of the staining fluid in a watch-glass. In using methyl violet, gentian violet, or malachite green

---

[1] These colors may be obtained from W. König, Berlin, N. W. Dorotheenstrasse, 35.

in a concentrated watery solution ¼-1 minute is long enough to color ; in the case of fuchsin and methyl blue it is well to use more diluted solutions for several minutes until the proper tinge is obtained. This latter has the advantage over the other stains in not over-coloring, but in staining the nuclei and bacteria distinctly and causing no precipitate (Ehrlich). Vesuvin (Bismarck brown) should be used in a concentrated watery solution for several minutes.

When the preparation is sufficiently colored it should be washed off carefully with water. Then the cover-glass is pressed between folds of filter paper and finally dried by holding it over the flame, and then examined in Canada balsam (dissolved in turpentine) or in cedar oil.

The fundamental principle in examining stained bac-teria preparation is to remove the diaphragm from the microscope stage, and, if possible, then use the Abbé illuminating apparatus [condenser] without the dia-phragm, making the contour of the preparation more indistinct, and thus causing the colored objects to be more prominent. But in all other microscopical exami-nations in which the endeavor is to have the clearest possible outline in an uncolored preparation—*e. g.*, in looking for hyaline casts, the narrowest diaphragm ad-missible should be used.

Almost all micro-organisms except the tubercle bacil-lus may be colored in dry preparation according to the above methods.

The staining of tubercle bacilli as done according to Ehr-lich. Aniline water is prepared by shaking up one or more ccm. of aniline oil with 20 ccm. [5 drachms] of dis-stilled water, and allowing it to stand a short time, and then filtering. To the clear filtrate, which may be heated to boiling in a test-tube to hasten the coloring, 5–10 drops of a concentrated alcoholic solution of diamond fuchsin are added in a watch-glass until the fluid begins to opalesce.

Instead of this solution, which should be prepared fresh every time, the following of Weigert-Koch may be used, which can be kept 10–12 days. Saturated aniline water 100 ccm. [3 ounces], a concentrated alcoholic solution of fuchsin or methyl violet 11 ccm. [2¾ drachms], absolute alcohol 10 ccm. [2½ drachms].

The preparations, smeared on a cover-glass, are allowed to float on the solution 3–12 hours (if the solution be heated, 5–20 minutes are long enough), then taken out with the forceps, dipped for a few seconds into diluted nitric or hydrochloric acid (1:3 water), then at once thoroughly washed with water. If the preparations have a red color the procedure should be repeated until this color disappears. All bacteria are decolorized by the acid except the tubercle bacillus (and lepra bacillus). This preparation should then be colored by a drop of concentrated watery solution of malachite green or methyl blue, again washed thoroughly with water, dried, and examined in cedar oil or Canada balsam. The tubercle bacilli will then be found to be colored red, while every thing else present is green or blue. The tubercle bacilli may be recognized with a power of 350 diameters.

The isolated method of staining micro-organisms according to Gram. The preparations are first colored for 1–3 minutes in a solution of aniline water which has been saturated with gentian violet, and then put into a solution of iodine in iodide of potash,[1] and then in absolute

---

[1] Iodine 1.0 [15 grains].
Iodide of potash 2.0 [30 grains].
Distilled water 300.0 [9½ ounces].

alcohol until the preparations are decolorized. The preparation is then to be stained with vesuvin and examined
in water, or dried and examined in cedar oil or Canada
balsam. The micro-organisms are colored bluish black.

In order to stain the pneumonococci of Friedländer,
and their capsules, there may be made, either a solution
of gentian violet in aniline water, or the solution which
Ehrlich uses to stain the plasma cells.[1] The preparation
should remain twenty-four hours in the solution, and
then put in 1% acetic acid for a few minutes, then in
alcohol, turpentine, and Canada balsam. The *Lepra
bacilli* are colored just as well in gentian violet or methyl
violet as, according to the procedure of coloring, the
bacilli of tuberculosis. The micro-organisms of *typhus,
recurrent fever, glanders, anthrax, pyæmia, erysipelas*, etc.,
may be shown with any of the basic aniline colors.

To color the gonococci, a drop of gonorrhœal pus is
pressed between two cover-glasses, spread out to a thin
film upon a slide, and to it are added a few drops of a
concentrated watery solution of methyl blue, which is
washed off in a half minute, and then dried and examined in Canada balsam or cedar oil.

The *actinomycosis* or radiating fungus, whose place
among the micro-organisms is doubtful, is found in pus
in the form of yellow-white granules of the size of a
millet seed, which consist microscopically of a large
number of fine radially arranged filaments, which end
in thick, shining knobs. The masses of actinomycosis
are often calcified, and should first be decalcified with
diluted hydrochloric acid. Staining is superfluous.

---

[1] Concentrated alcoholic solution of gentian violet, 50.0 [1½ ounces].
Glacial acetic acid, 10.0 [2½ drachms].
Distilled water, 100.0 [3 ounces].

CHAPTER XII.

THE NERVOUS SYSTEM.

Testing the Sensibility.

WE distinguish *Anæsthesia*, a loss or diminution of sensation. *Hyperæsthesia*, an exaltation of the same, weak stimuli causing unpleasant sensations. *Paræsthesiæ*, or abnormal sensations which are not due to external causes, as itching, crawling, formication, furry feeling, abnormal sensation of heat and cold. *Neuralgiæ* are attacks of pain which are confined to a certain nerve region, and they generally follow the course of the nerve. They are generally increased by pressure on the nerve on that part which is subcutaneous, and when it is pressed against a bone (pressure point). In genuine neuralgia the pain is in paroxysms.

The *sensibility* may be equally diminished for all kinds of sensation or for some kinds only (total and partial anæsthesia). These kinds of sensations are :

Touch sense, which may be tested by delicately touching the part affected with the finger-tip or any other object. The patient, whose eyes are covered, should be very attentive to note the slightest touch. The test may be made between smooth and rough (woollen) objects. The temperature sense must then be excluded.

The *sense of locality*.—The patient should be touched,

and then asked to show what part was touched. Healthy individuals rarely miss, or err by 1–2 cm. [$\frac{1}{2}$–$\frac{3}{4}$ inch] only. Or, by using a pair of compasses, the smallest distance may be found in which the two points applied at the same time and in the same way can be recognized as two points. The distance in health for the following localities is as follows :

Tip of the tongue	1 mm.	[$\frac{1}{40}$ inch].
Tip of the finger . .	2 "	[$\frac{1}{20}$ "].
Red surface of the lips . .	3 "	[$\frac{1}{13}$ "].
Dorsal surface of the first and second phalanx and inner surface of the fingers	6 "	[$\frac{1}{7}$ "].
Tip of the nose . . .	7 "	[$\frac{1}{6}$ "].
Dorsal and palmar surfaces	8 "	[$\frac{1}{5}$ "].
Chin	9 "	[$\frac{1}{4}$ "].
End of big toe, cheek, and eyelids	12 "	[$\frac{1}{3}$ "].
Bridge of the nose . .	13 "	[$\frac{1}{3}$ "].
Heel . . .	22 "	[$\frac{1}{2}$ "].
Back of the hand	30 "	[$\frac{3}{4}$ "].
Neck	35 "	[$\frac{7}{8}$ "].
Forearm, leg, dorsum of the foot .	40 mm.	[1 inch].
Back	60–80 mm.	[1$\frac{1}{2}$–2 inches].
Upper arm and thigh . .	80 mm.	[2 "].

Sense of pressure (muscular sense).—The extremity to be tested should be firmly supported and weights laid upon it, a small piece of board being put between the weight and the extremity to eliminate the sense of temperature. Under normal conditions a difference of $\frac{1}{10}$ of the original weight can be recognized, as well as a minimum pressure of 0.002 to 1.0 gram [$\frac{1}{32}$ to 15 grains]. Greater disturbances of the muscular sense can be recognized by pressure with the finger.

Sense of temperature.—Test tubes, or metal vessels filled with water of different temperature, are applied to the skin. Between 25–35° C. [77–95° F.] a difference of $\frac{1}{2}$° in the temperature is recognized by a healthy person. The test may be made by letting the patient endeavor to distinguish between warm breath near the skin and cold breath from a distance. Many patients will say that the irritation from the cold (ice) is hot, and *vice versa* (perverted temperature-sense).

Electro-cutaneous sensibility. — By applying a metal brush to the skin, it may be ascertained with what strength of current (distance of the coils) the faradic stream has been felt.

Sensibility to pain is tested by sticking with a needle, pinching, pulling the hair, and using strong electric current. If strong and painful irritation, as deep puncture with a needle, be felt as if the needle only touched the skin, without pain, then it is called *analgesia.* Analgesia occurs with unimpaired tactile sense in hysteria and tabes. Also the reverse may be noticed, *i. e.*, simple contact may cause pain. There is often a delay experienced in the transmission of the sensation of pain, or an abnormal after-sensation, and, at times, the tactile and pain-producing sensations are separated and are perceived one after the other (a double sensation).

The *sensitiveness of the deep parts*—the muscles, fasciæ, tendons, ligaments, joints, periosteum, and bones—is classed as follows :

1. The ability to judge of the weight of a body when raised up, *i. e.*, the sense of force ; this is tested by lifting up a cloth to which weights are gradually added, and estimating the weight. The sense of force is finer than the sense of pressure.

2. The ability to judge, with closed eyes, of the position of one's extremities and their passive movements ; or it may be tested by letting the patient close his eyes and attempt to touch one extremity with the other ; or, further, the power to hold the body in an upright position when the eyes are closed. If the patient stand firm with open eyes, and totter or fall when the eyes are closed (symptom of Romberg), then the sensibility of the limb is diminished.

Testing the Motility.

When the power of voluntary motion in a muscle is completely lost, we speak of *paralysis,* and when this is only weakened, of *paresis.* According to the extent of the paralysis, we speak of a *monoplegia* or paralysis of single muscles or ·group of muscles, or of an extremity by itself ; *hemiplegia,* paralysis of one side ; *paraplegia,* paralysis of corresponding parts of both sides of the body, *i. e.,* of both limbs, or of both arms, or of all four extremities. ·

It should be noticed whether the state of muscular tension deviates from the normal or not. In *diminished tension* the paralyzed muscles are relaxed, and make no opposition to passive movements. This relaxed paralysis occurs principally in peripheral lesions in diseased conditions of the anterior gray horns or gray nuclei, as in infantile paralysis.

In *increased tension* the muscles are stiff, rigid, and oppose all passive movements. If the tension is increased, it may lead to contraction. The rigid, so-called spastic paralysis presupposes a *central* lesion (of the brain or spinal cord), and occurs principally in degeneration of the lateral pyramidal tracts of the spinal cord, in spastic spinal paralysis, in amyotrophic lateral sclerosis, in cerebral apoplexy or embolisms, etc. Spastic paralysis goes hand in hand with an *exaggerated tendon reflex*, while in relaxed paralysis there is a diminution or absence of the tendon reflex.

In contradistinction to the *organic* paralyses, in which the motor tract in any part is injured, we speak of a *functional* paralysis, where the motor tract is unimpaired, *i. e.*, in hysterical paralysis.

Ataxia means the inability, with intact power, to coördinate the separate muscles to a certain action—that is, a condition in which the patient makes clumsy motions, when he was previously skilful. He is asked to quickly reach for a certain object, to button a button, to write, to walk a straight line, to turn himself around, to describe a circle with the foot, etc. The patient shows ataxia when on the feet, by standing with the feet wide apart, and walking stiff-legged or stamping along. When the patient cannot control his movements with his eyes (in the dark, and with closed eyes), the ataxia is generally worse. Ataxia occurs in diseases of the spinal cord (tabes), as well as of the brain and peripheral nerves (cerebellar, alcoholic, diphtheritic ataxia).

Motor Symptoms of Irritation.

Spasms, or involuntary muscular movements, are divided into *clonic* (interrupted quiverings of short duration), and *tonic* (contractions of longer duration). If the tonic spasms extend to most of the muscles, it is called *tetanus. Convulsions* are numerous quick, powerful clonic spasms, especially if they extend over the whole body.

Tremors occur either in muscles at rest (in paralysis

agitans), or in muscles voluntarily moved, especially in the movements which demand strength or precision (tremor of intention as in multiple sclerosis). Trembling of the eyes is called *nystagmus* (multiple sclerosis).

Choreic movements are quick, involuntary, and incoördinate movements which interrupt and prevent the voluntary motion. They occur in chorea minor, and at times, on one side after (or before) hemiplegia.

Further, the following are to be mentioned : *Compulsory movements* (riding motion), *accompanying motions* (generally central), *athetosis motions* (slow and rhythmic exaggerated movements of the hands), and *cataleptic* (waxy) *muscular rigidity*.

Diagnosis by Means of Electricity.

The test should be made both with the *faradic* (interrupted) and *galvanic* (constant) current, both by *direct* application to the muscles, or by *indirect* excitation of the muscles through the nerves. The *indifferent* [*i. e.*, non-active] pole (a long, flat electrode) is placed on the sternum, and the other, *different* [*i. e.*, active] pole, on the nerve or muscle to be examined. A small, button-shaped electrode serves for the different pole, since it must be taken into account, for the effect of the electrical excitation, that the current reach the part to be excited, with the greatest *density*. The *density* (D) is greater in proportion as the *intensity* (I) of the current is greater, and the *section* of the conductor (S) is smaller at the spot : $D = \frac{1}{s}$.

The electrodes, as well as the skin of the patient, should be well moistened with warm water. The situation of the points in which a muscle or nerve may be excited, is shown in the accompanying illustrations.[1] By gradually increasing the strength of the current, we arrive at a point where the first minimum muscular contraction takes place.

The examination is begun with the faradic current and generally with the current of the secondary coil.

Fig. 55.

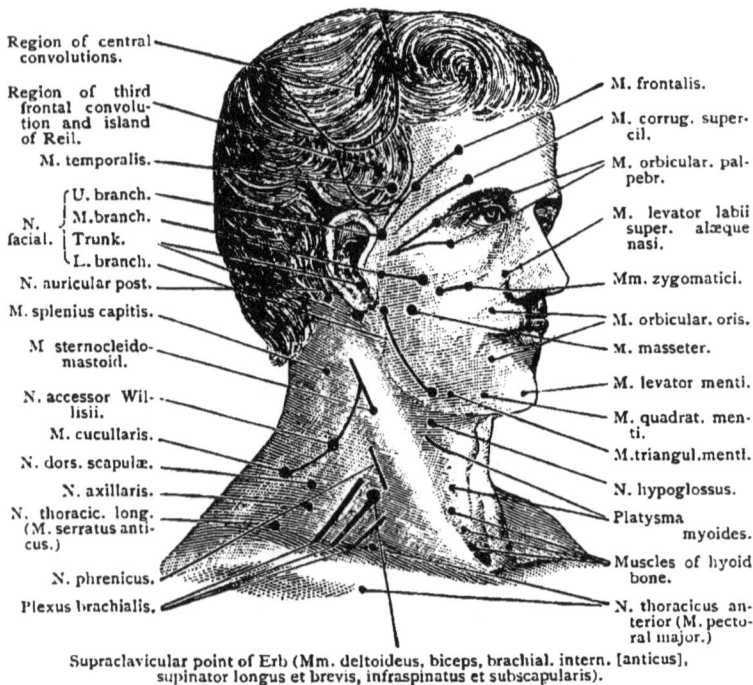

Region of central convolutions.

Region of third frontal convolution and island of Reil.

M. temporalis.

N. facial. { U. branch. M.branch. Trunk. L. branch.

N. auricular post.

M. splenius capitis.

M sternocleidomastoid.

N. accessor Willisii.

M. cucullaris.

N. dors. scapulæ.

N. axillaris.

N. thoracic. long. (M. serratus anticus.)

N. phrenicus.

Plexus brachialis.

M. frontalis.

M. corrug. supercil.

M. orbicular. palpebr.

M. levator labii super. alæque nasi.

Mm. zygomatici.

M. orbicular. oris.

M. masseter.

M. levator menti.

M. quadrat. menti.

M.triangul.mentl.

N. hypoglossus.

Platysma myoides.

Muscles of hyoid bone.

N. thoracicus anterior (M. pectoral major.)

Supraclavicular point of Erb (Mm. deltoideus, biceps, brachial. intern. [anticus], supinator longus et brevis, infraspinatus et subscapularis).

As a standard of measurement of the intensity of the current, the distance between the coils (R. A.) is expressed in millimetres, and the current is *stronger* in proportion as the coils are *further apart.*

----

[1] These are drawn from the illustrations in the text-books of Ziemssen, Erb, Bernhardt, Rosenthal and Eichhorst. For details as to electro-diagnosis, as well as electro-therapeutics, see these books.

Also the faradic current must be graded by moving the iron rod (in the coil) ; for the current is so much the *stronger* in proportion as the rod is pushed *further* into the primary spiral.

In using the galvanic current[1] the cathode[2] (negative zinc pole) is applied to the muscle or nerve to be examined. By gradually increasing the strength of the current, it may be determined what the least intensity is, with which, at the closing of the current, a contraction takes place (cathodal closing contraction, KaSZ[3]). The intensity is noted by giving the number of elements used, or by reading off the number on the galvanometer.

Then the current is used unclosed, with the commutator, (from N, the normal position, to W, change), by which the exciting electrode becomes the anode (the positive, carbon, or copper pole), and determines the minimum of contraction on closing (anodal closing contraction, AnSZ) and on opening (anodal opening contraction, AnOZ). The closing and opening of the

---

[1] In filling the carbon zinc elements, the following ·fluid is used :
 Bichromate of potash, 70.0 [$17\frac{1}{2}$ drachms].
 Water, 900.0 [28 ounces].
 Concentrated sulphuric acid, 170.0 [$5\frac{1}{2}$ ounces.]
 Sulphate of mercury, 10.0 [$2\frac{1}{4}$ drachms].
The last ingredient is to keep the zinc amalgamated.

[2] In order to distinguish the two poles, the ends of the wire should be immersed in a solution of iodide of potash and starch, and blue clouds, due to the free iodine and starch, are formed at the *anode.* Or the ends of the wires may be dipped in water, and the bubbles of hydrogen will show which is *cathode*, while the anode is recognized by the absence of bubbles, due to the rapid oxidation of the oxygen as fast as it is formed.

[3] [For the sake of uniformity, the German abbreviations are used throughout.]

current should be effected by the interrupter without changing the position of the electrodes.

Under normal conditions, the results of the irritation, on gradually increasing the intensity of the current, are in the following order :—

(1) Cathodal closing contraction, KaSZ.

(2) Anodal opening contraction, AnOZ.

(3) Anodal closing contraction, AnSZ.

(4) Cathodal closing tetanus, KaOZ (lasting contraction with KaS).

(5) Cathodal opening contraction, KaOZ.

This law holds good, however, only in indirect irritation of the nerves. In direct application of the electrode to the muscle, there are generally closing contractions, and AnSZ may be equal to KSZ or even greater than it.

The contractions are short, quick, and may be excited through the nerve or muscle.

The intensity of the current is expressed by the number of elements used, or still better, when an absolute galvanometer is at hand, in milliampères.

According to Ohm's law $I = \frac{E}{W}$; that is, the strength of the current or intensity (I) is in proportion to the electro-motive force (E, number of elements), and is in inverse proportion to the whole amount of the resistance present in the electric current. Now an ampère is that strength of current (I) which is generated by the electro-motive force (E) of 1 volt in an electric current of resistance (W), of 1 ohm. An ampére then, is equal to $\frac{1 \text{ volt}}{1 \text{ ohm}}$. One volt is equal to $\frac{9}{10}$ of the electromotive force of a Daniell element ; one ohm is equal to a column of mercury 106 cm. long, and 1 square millimetre in section (1.06 Siemen's unit). For medical purposes, no strength of current higher than 20 thousandth (milli-) ampères is used. With motor nerves superficially situated KaSZ occurs normally with currents of 1–3 MA strength. The strength of the current may be varied, either by inserting more or less elements, or by means of a rheostat, by which resistance of different degrees may be inserted into the current.

The *resistance* in the dry epidermis is at first very great, but after using the galvanic current for some time, and thoroughly moistening the skin, the resistance is considerably diminished, so that by using a current of medium strength with the number of elements (E), remaining the same, the strength of the current (I) increases to a certain point. A current which is not felt at the beginning of the examination, and causes no contraction, may, by keeping the current closed, and diminishing the resistance without changing the number of elements, so increase as to become painful and cause evident contraction.

Quantitative changes in the electro-irritability, that is simple increase or diminution, are judged by comparing both sides of the body (in unilateral affections), and by testing analogous points which have approximately the same irritability in health, *e. g.*, the frontal nerve, the spinal accessory in the neck, the ulnar nerve above the olecranon, and the peroneal nerve between the bend of the knee and the head of the fibula (Erb). Here it should not be forgotten that the conductive resistance of the skin is different in different parts of the body, and in different individuals.

Simple increase of electric irritability occurs in tetanus, and a *simple diminution* of electric irritability may develop in all paralyses of long duration, which begin with simple non-degenerative muscular atrophy, *e. g.*, after apoplexy and muscular atrophy of the joint troubles.

When there is a very great diminution of the electric irritability, a contraction may be caused, with the strongest currents, with

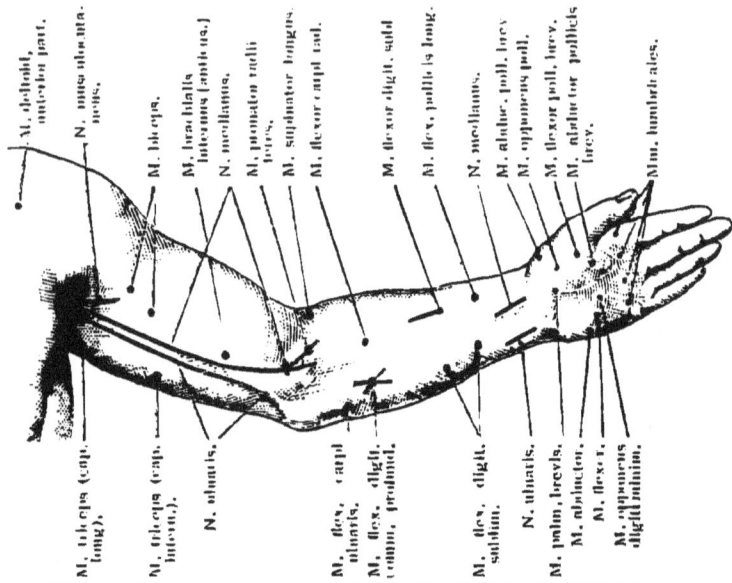

Fig. 57.

M. deltoid, anterior part. — N. musculocutaneous. — M. biceps. — M. brachialis internus [anticus.] — N. medianus. — M. pronator radii teres. — M. supinator longus. — M. flexor carpi rad. — M. flexor digit. subl. — M. flex. pollicis long. — N. medianus. — M. abduc. poll. brev. — M. opponens poll. — M. flexor poll. brev. — M. abductor pollicis brev. — Mm. lumbricales.

M. triceps (cap. long). — M. triceps (cap. intern.). — N. ulnaris. — M. flex. carpi ulnaris. — M. flex. digit. comm. (profund.) — M. flex. digit. sublim. — N. ulnaris. — M. palm. brevis. — M. abductor. — M. flexor. — M. opponens digiti minimi.

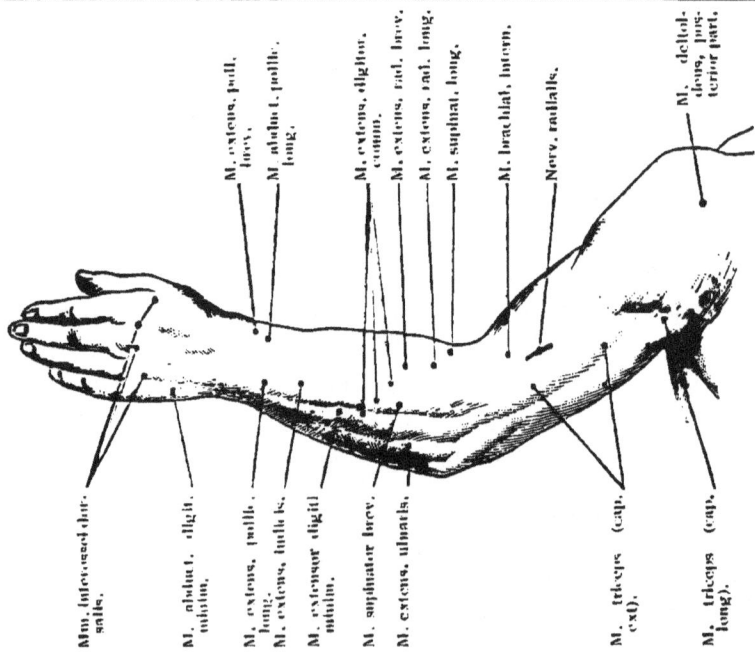

Fig. 56.

M. extens. poll. brev. — M. abduct. pollic. long. — M. extens. digitor. comm. — M. extens. rad. brev. — M. extens. rad. long. — M. supinat. long. — M. brachial. intern. — Nerv. radialis. — M. deltoideus, posterior part.

Mm. interossei dorsales. — M. abduct. digit. minimi. — M. extens. pollic. long. — M. extens. pollic. br. — M. extensor digiti minimi. — M. supinator brev. — M. extens. ulnaris. — M. triceps (cap. ext). — M. triceps (cap. long).

118

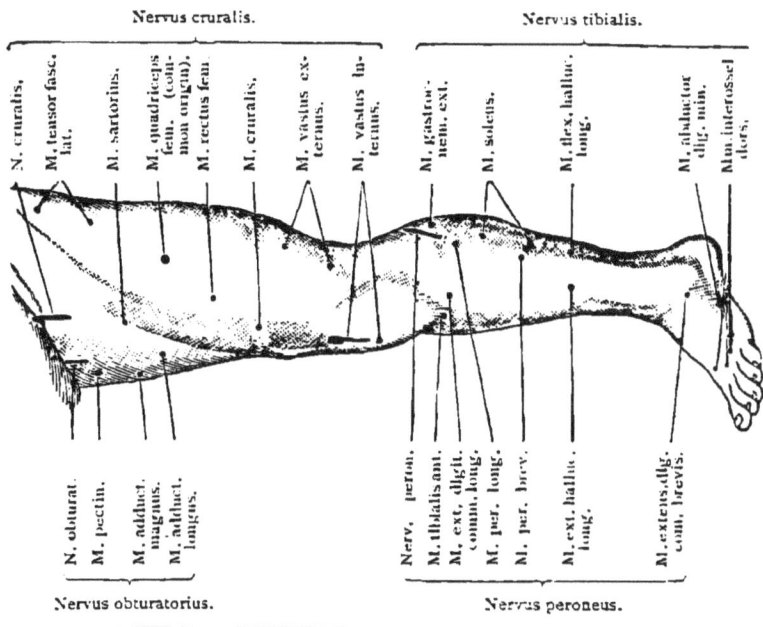

Fig. 59.

(top labels) N. cruralis. · M. tensor fasc. lat. · M. sartorius. · M. quadriceps fem. (common origin). · M. rectus fem. · M. cruralis. · M. vastus externus. · M. vastus internus. · M. gastrocnem. ext. · M. soleus. · M. flex. halluc. long. · M. abductor dig. min. · Mm. interossei dors.

(bottom labels) N. obturat. · M. pectin. · M. adduct. magnus. · M. adduct. longus. · Nerv. peron. · M. tibialis ant. · M. ext. digit. comm. long. · M. per. long. · M. per. brev. · M. ext. halluc. long. · M. extens. dig. comm. brevis.

Nervus obturatorius.   Nervus peroneus.

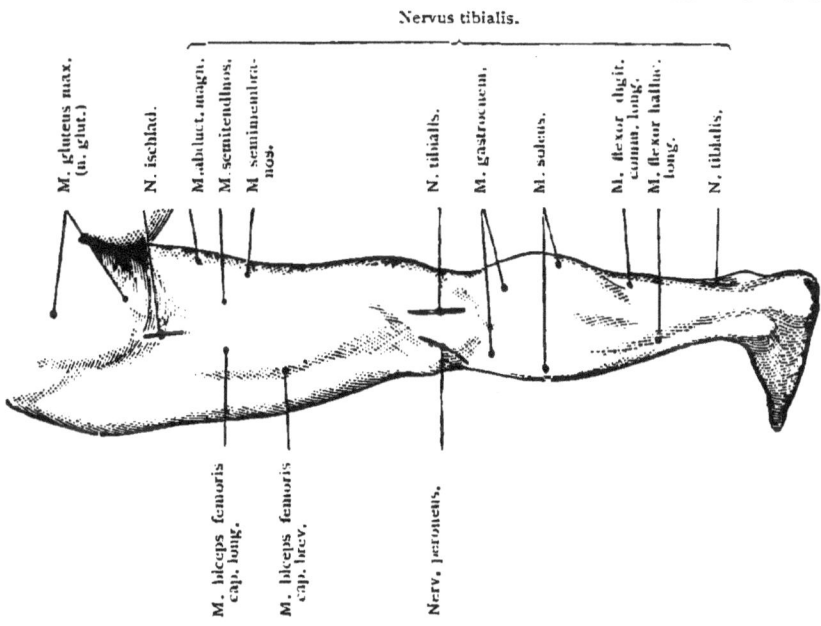

Fig. 58.

Nervus tibialis.

(top labels) M. gluteus max. (m. glut.) · N. ischiad. · M. abduct. magn. · M. semitendinos. · M. semimembranos. · N. tibialis. · M. gastrocnem. · M. soleus. · M. flexor digit. comm. long. · M. flexor halluc. long. · N. tibialis.

(bottom labels) M. biceps femoris cap. long. · M. biceps femoris cap. brev. · Nerv. peroneus.

closed connection only, from the anode to the cathode (Volta's alternative), and even this may be entirely extinguished.

A qualitative change in the electric irritability, is called the *reaction of degeneration* (EaR). When a motor nerve is diseased or cut off peripherally from its trophic centre (the anterior horns of the spinal cord, or the gray matter of the cervical nerves), or if the trophic centre itself be diseased, a motor paralysis appears, the nerve becomes degenerated, and the degeneration (degenerative atrophy) reaches to the muscle supplied by it. The electric irritability of the nerve diminishes for the *faradic*, as well as for the *galvanic* current, and is destroyed after about two weeks, *i. e.*, the nerve ceases to conduct the electric current as well as the will. Also the direct faradic irritability of the muscle is diminished and disappears ; on the contrary, in the 2d or 3d week there is an *increase of the direct muscular irritability* for the *galvanic* current, the contractions occur with the weakest current, but *are long drawn out and slow*, and the formula of contraction is changed. The AnSZ occurs with the same or weaker current, as the KaSZ, and the KaOZ becomes like the AnOZ. In one or two months the galvano-muscular irritability diminishes, and disappears in a few months. If recovery take place the muscular tonus and voluntary motion appear, but the electric irritability returns only gradually to the normal.

This *" complete reaction of degeneration "* only occurs in grave lesion of the nerves (transverse rupture, severe rheumatic facial paralysis); when the conditions of degeneration are not so grave, there is often an incomplete, or even no *"partial reaction of degeneration."* In the latter case, the irritability for the nerves is retained, and also the direct faradic muscular irritability. In di-

rect galvanic irritation of the muscles, there are neverthe-
less, hyperexcitability, and change of the formula of
contraction (AnSZ > KSZ), and a *slow contraction.*
The latter is the actual characteristic of EaR.

The reaction of degeneration is present in *peripheral lesions of the
motor nerves,* of a traumatic, rheumatic, neurotic, or diphtheritic
nature, also in disease of the gray matter of the anterior horns of
the spinal cord, and of the gray nuclei of the medulla, *e. g.,* in infantile paralysis and lead paralysis; also sometimes in progressive
muscular atrophy, bulbar paralysis, amyotrophic lateral sclerosis,
myelitis, etc.

The EaR is absent, however, in all cerebral paralyses (apoplexies),
and in those spinal paralyses which have a central cause from the
trophic centre, and also in pure myopathic paralysis (pseudo-hyper-
trophy of the muscles).

The *trophic behavior* of the paralyzed muscles is entirely analogous to the electric behavior. In disease of
the gray matter of the anterior horns, as well as in lesions
of the motor nerves peripherally from themselves, a
degenerative atrophy occurs, while in paralysis whose
cause lies in the motor tract central from the gray matter of the anterior horns, a slight atrophy of the paralyzed muscle takes place, but not until after a long time
(atrophy of inactivity).

In degenerative atrophy of the muscles, there are often
fibrillary contractions observed in them.

Reflexes.

We distinguish skin (superficial) and tendon (deep)
reflexes. It is not certain whether the latter are actually
reflex or not. They do not behave alike, and may be
often completely different.

Among the **skin reflexes** which are more or less
present in health, are :

Reflex of the sole of the foot :   In exciting the sole of the foot by tickling, stroking, sticking, touching it with ice, there is dorsal flexion of the foot, and when the irritation is strong, the leg is drawn up against the body.

Cremaster reflex : In exciting the inner surface of the thigh, the corresponding testicle rises up.

Reflex of the abdominal walls, gluteal and scapular regions :   In irritating the skin in these regions, the corresponding muscles contract.

Tendon Reflexes.

Patellar reflex :   If the patellar tendon be percussed while the leg is crossed and *completely relaxed,* and the patient's attention be withdrawn, there is a contraction of the quadriceps and the leg is extended.

Patellar clonus : If the patella be pushed quickly down and held there firmly, there is a rhythmic contraction of the quadriceps.

Reflex of the tendo Achillis :   In percussing the tendo Achillis, there is caused a contraction of the calf muscles.

Foot clonus :   If the foot be seized by the ball of the great toe, and be pressed quickly upward while the knee is slightly bent, there is a rhythmic plantar-contraction of the calf muscles.

In health the patellar reflex is constant, and the reflex of the tendo Achillis frequent.   The presence of the remaining tendon reflexes, also of the upper extremities (biceps, triceps, flexors of the hand etc.), is considered as a diseased reflex irritability.

In order that the conditions of reflex tendon may occur, it is, above all things, necessary that the *reflex circuit* be entire.   This reflex circuit is formed by the sensory nerve tracts, which go from the muscle or tendon or fascia, to the spinal cord and motor tracts which descend to the muscles, as well as to that part of the spinal cord connecting both. The reflexes are *extinguished* when the reflex circuit is interrupted in any part of its course, *i. e.,* when the centripedal or centrifugal nerves, or their connection with the spinal cord (Burdach's pyramidal gray substance), are injured. The reflexes are *increased* when these nerves are in an abnormally excited condition, or when the inhibitory fibres are interrupted in their course from the cerebrum through the lateral tracts to the reflex circuit.

The tendon reflex is *extinguished* in polyomyelitis anterior (infantile paralysis), tabes dorsalis, peripheral nerve lesions, and diffuse myelitis. *Increase* of the tendon reflex is observed in sclerosis of the lateral tracts, in amyotrophic lateral sclerosis, in hemiplegia with contraction, in multiple sclerosis, division of the spinal cord above the reflex circuit, and further in dementia paralytica and hysteria.

Paradox contraction (Westphal) : When the foot of the recumbent patient is quickly and firmly flexed, there is sometimes a contraction of the tibialis anticus, its tendon is prominent, and the foot remains for a short time in this position after it has been let go.

Among the reflex functions may be mentioned the passing of urine and fæces, the *sexual reflex* and *pupil reflex.*

The *pupil* is supplied with fibres from the oculomotorius for the sphincter pupillæ as well as those from the sympathetic for the dilatator. A centre for the pupil reflex lies in the lower cervical region (cilio-spinal centre). Irritation of the *sympathetic fibres* coming from the centre, causes dilatation of the pupil (mydriasis spastica), paralysis of the same, a narrowing of the pupil (myosis paralytica). Irritation of the *oculomotorius*, on the contrary, causes narrowing of the pupil, and paralysis of it, dilatation, and want of reflex for light as well as for accommodation to near objects. Reflex rigidity of the pupil for light, with retained movement for accommodation for near objects, occurs with narrowing and inequality of the pupils, most frequently in tabes dorsalis and dementia paralytica.

The Most Important Clinical Points in the Anatomy of the Nervous System.

Brain and Spinal Cord.—The psychomotor region of the cerebral cortex is formed by the two central convolutions and their connecting part on the median surface, the lobus paracentralis. The centre for the leg probably lies in the latter and in the two upper thirds of both the central convolutions ; and in the middle third of the anterior convolution lies the centre for the arm ; and in the lower third of the anterior convolution, the centre for the face (facialis, hypoglossus). Bordering on the latter, and in the posterior portion of the third left lower frontal convolution, as well as in the island of Reil, lies the centre for speech. When this is injured, aphasia occurs. The cortex of the parietal lobes is brought into relation with the sensible tract ; the occipital lobe is the cortical centre for the sense of sight. As there is only a partial crossing of the fibres in the chiasma of the optic nerve, an injury to the occipital lobe or to the optic tract as far as the chiasma may cause hemianopsia, *i. e.*, blinding on the corresponding side of the retina ; thus a diseased condition of the right side causes a blindness in the left half of the field of vision. A diseased condition of the central part of the chiasma causes a bitemporal hemianopsia. In an injury to the optic nerve beyond the chiasma, there occurs amblyopia, or amaurosis of one entire eye. The temporal lobes are connected with the sense of hearing. The functions of the corpus striatum, nucleus lentiformus, and thalamus opticus are not exactly known, but as they border on the inner capsules, diseased conditions of them may indirectly cause hemiplegia. The cerebellum is said to be the centre for coördination. When it is diseased, we have ataxia, dizziness, and vomiting.

The motor fibres of the psychomotor cortical portion pass through the peduncle of the corpus callosum, and converge to the inner capsule, where they run in the middle third of the posterior crus between the opticus thalamus and the nucleus lentiformis. In its posterior third the sensory tracts are found. From the inner capsule the motor tracts pass through the pes cruris cerebri (the sensory through the tegmentum cruris cerebri) into the pons.

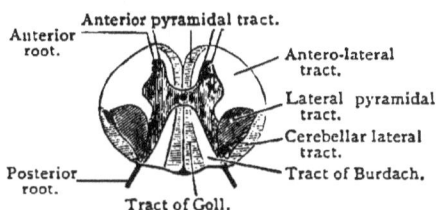

Anterior pyramidal tract.
Anterior root.
Antero-lateral tract.
Lateral pyramidal tract.
Cerebellar lateral tract.
Posterior root.
Tract of Burdach.
Tract of Goll.

The motor fibres, after their exit from the pons into the medulla, form the pyramidal bodies, and here, for the most part, cross. The crossed fibres in the lateral tract of the spinal cord pass downwards (lateral pyramidal tract, Fig. 60); only a small part of the motor fibres remains uncrossed, and passes down in the middle part of the anterior tract (anterior pyramidal tract). A destruction of any part of this motor tract produces, not only a paralysis of the muscle concerned, but also descending degeneration of the pyramidal tracts, inasmuch as their trophic centre is situated in the cerebrum. The motor fibres pass out of the pyramidal tracts into the anterior cornu of the gray matter, whose great ganglion cells form the trophic centre for the peripheral motor nerves and the muscles ; and from there they pass through the anterior roots to the periphery of the body.

Injuries of the motor nerves beyond the gray anterior cornua, or a morbid condition of them, produce a degeneration of the nerves, as well as paralysis and atrophy of the muscles, with the reaction of degeneration. These paralyses are characterized as *peripheral*, in distinction from *central paralyses*, which are caused by a lesion of the motor tracts proximal to the gray anterior cornua.

Inasmuch as in the cerebral cortex the motor centres of the single muscular regions lie far apart from each other, a lesion in that particular place generally produces *monoplegia*, *i. e.*, a paralysis of one member or of one group of muscles alone, which is often connected with paroxysmal cramps in the paralyzed portion (cortical epilepsy, Jackson's epilepsy). Lesion of the inner capsule generally produces total hemiplegia, as well as affections of the crus cerebri and the pons. (In case of diseased condition of the crus cerebri, there is often with it a crossed paralysis of the oculomotorius ; when the pons is diseased, there is crossed paralysis of the facialis.) All these paralyses affect the opposite side of the body, while affections of the spinal cord *before* the pyramidal decussation cause paralyses of the same side. As most lesions, however, affect the spinal cord *on both sides alike*, paraplegia is the principal form of spinal paralysis (myelitis, compression of the spinal cord by spondylitis or tumors). Injuries of the anterior roots, of the plexus, and of the nerves, produce paralyses of single groups of muscles.

The sensory nerves, whose trophic centre is situated in the intervertebral ganglia, enter the spinal cord through the posterior roots, cross shortly after their entrance (deep decussation), and ascend through the *posterior columns* to the brain, so that the inner columns (of Goll) contain the long, ascending bundles of fibres, while the outer columns (of Burdach) are made up of short bundles which run to the gray posterior cornua. Besides that, long bundles in the lateral column of the cerebellum pass upward.

In cases of *transverse section* of the spinal cord, the columns of Goll and cerebellar lateral columns degenerate upward from the point of injury, and the pyramidal column downward. In case of a *lesion of one side* of the spinal cord, there appears a motor paralysis of the same side, and anæsthesia of the other side ; besides this, a narrow anæsthetic belt around the body at the height of the lesion (Brown-Séquard). In case of tabes dorsalis the posterior columns are diseased, and in spastic paralyses the lateral columns (tabes spastica). In amyotrophic lateral sclerosis the anterior cornua and lateral columns are diseased, and in infantile paralysis and progressive muscular atrophy the gray anterior cornua are diseased. In diseases of the gray nuclei of the medulla oblongata (bulbar paralysis) there are disturbances of speech and deglutition caused by paralysis and atrophy of the lips, soft palate, muscles of deglutition, and larynx.

Cranial Nerves.

1. *Olfactorius.* The testing of the sense of smell is accomplished by holding before the nose odoriferous and irritating substances, such as volatile oils, asafœtida, musk, etc.

2. *Opticus.* Test the sharpness and field of vision, and sense of color, and then examine with the ophthalmoscope.

3. *Oculomotorius* supplies the levator palpebræ superioris, rectus superior, internus, and inferior, obliquus inferior, and sphincter pupillæ. In paralysis, there is ptosis, diplopia, dilatation, and absence of pupillary reaction, and disturbance of accommodation,

4. *Trochlearis* supplies the obliquus superior.

5. *Trigeminus.* The motor part supplies the muscles of mastication, the masseter, temporalis, pterygoid, mylohyoid, and the anterior belly of the biventer. The sensory part supplies the skin of the face and head as far as the ears. The first branch goes to the skin of the forehead, of the top of the head, of the upper eyelids, and of the bridge of the nose. The second branch supplies the upper half of the cheek and upper lip, and the third branch, the lower half of the cheek, the skin in the temporal region, and the chin. Besides this, the trigeminus supplies the cornea and conjunctiva and the mucous membrane of the mouth and nose, and the dura mater with sensory fibres. The lingualis from the trigeminus is the nerve of taste for the anterior two thirds of the tongue.

6. The *abducens* supplies the abducens muscle. When it is paralyzed, the eyeball cannot be turned outward.

7. The *facialis* supplies all the mimic muscles of the face, and the stylohyoideus, and the posterior part of the biventer. From the relation of the facialis to the petrosus superficialis major nerve, and to the chorda tympani, it is clear that in a lesion of this nerve, proximal to the ganglion geniculi, the soft palate on the same side is paralyzed and hangs lower, and that in a lesion between the ganglion geniculi and the passage of the chorda tympani, disturbances of taste occur in the anterior two thirds of the tongue, with decrease of the salivary secretion. In central paralysis of the facialis, only the lower half of the face is usually paralyzed ; in peripheral paralysis, only the upper part.

8. *Acusticus.* The power of hearing should be tested, and otoscopic examination made.

9. *Glossopharyngeus.* The nerve of taste for the posterior third of the tongue supplies the palate with sensory fibres. The test is to apply quinine, sugar, salt, or vinegar to the part.

10. The *vagus* supplies the larynx, pharynx, and œsophagus with motor and sensory fibres, and sends fibres to the contents of the chest and abdomen. Irritation of the vagus causes slowing of the pulse ; and paralysis of the nerve, a quickening of the pulse and slowing of the respiration.

11. The *accessorius* supplies the sterno-cleido-mastoid and the trapezius.

12. The *hypoglossus*, the motor nerve of the tongue, supplies the

genio-glossus, hyo-glossus, stylo-glossus, the innermost muscles of the tongue, the genio-hyoideus, omo-hyoideus, sterno-hyoideus, hyo-thyroideus, and sterno-thyroideus. In paralysis of the hypoglossus, the tongue turns toward the paralyzed side.

Spinal Nerves.

1. *Plexus cervicalis* (1st-4th cervical nerve) supplies the post-occipital region behind the ear, neck, and shoulders with sensory nerves ; the deep cervical muscles and the scaleni, with motor nerves. From the fourth cervical nerve the phrenic branches, and forms the motor nerve of the diaphragm.

2. *Plexus brachialis* (5th-8th cervical nerve, 1st and 2d dorsal nerve). In lesion of a certain part of this plexus, there is a motor paralysis of the deltoid, biceps, brachialis internus [anticus], supinator longus, infraspinatus (paralysis of Erb).

The *nervi thoracici anteriores* supply the musculus pectoralis major and minor.

The *nervus thoracicus longus* supplies the musculus serratus anticus major [serratus magnus].

The *nervus dorsalis scapulæ* supplies the musculi rhomboidei, levator [anguli] scapulæ, and serratus posticus superior.

The *nervus suprascapularis* supplies the musculus supraspinatus and infraspinatus.

The *nervus subscapularis* supplies the musculus subscapularis, teres major, and latissimus dorsi.

The *nervus axillaris* supplies the musculus deltoideus, teres minor, and sensory fibres go to the skin of the outer side of the upper arm.

The *nervus cutaneus medius* and *medialis* supply the skin of the median (inner surface) side of the forearm.

The *nervus musculocutaneus* supplies the musculus biceps, coraco-brachialis, brachialis internus [anticus], and the skin on the radial side of the forearm.

The *nervus medianus* supplies the musculus flexor carpi radialis, pronator [radii] teres and pronator quadratus, flexor digitorum communis superficialis and profundus (in part), palmaris longus, flexor pollicis longus and brevis, abductor and opponens pollicis ; the skin of the palmar surface of the hand from the thumb to the middle of the third [ring] finger ; and the dorsal side of the ungual phalanx of the first and second finger.

In paralysis of this nerve, pronation and flexion of the hand is almost entirely impossible, and flexion and opposition of the thumb and flexion of the finger in the last two phalanges, is impossible ; on the contrary, the first phalanges can be flexed by the interossei. With the last three fingers, whose flexor profundus is partly supplied by the nervus ulnaris, the power of grasping is still retained.

The *nervus ulnaris* supplies the musculus flexor carpi ulnaris, flexor digitorum profundus for the last three fingers, the muscles of the ulnar side of the hand, the interossei, lumbricales, adductor pollicis ; the skin of the ulnar side of the hand on the palmar side as far as the middle of the third [ring] finger, and on the dorsal side to the middle of the second finger.

In paralysis of this nerve there is diminished power of lateral movement towards the ulnar side as well as loss of power to flex the last three fingers : further, also, there is loss of motion of the little finger in flexion of the first phalanges and extension of the last phalanges of the four last fingers, and loss of power to spread the fingers out and draw them together. In paralyses which have existed a long time we have the claw-like position of the hand, in which case the first phalanges are flexed towards the dorsal surface and the end phalanges towards the palmar surface. This is caused by atrophy of the interossei.

The *nervus radialis* supplies the extensors of the arm, hand, and fingers, the musculus triceps, supinator longus and brevis, all the muscles on the posterior surface of the forearm, namely, the extensor carpi radialis longus and brevis, extensor carpi ulnaris, extensor digitorum communis, extensor indicis and digiti minimi, extensor pollicis longus and brevis, abductor pollicis longus. The cutaneous branches go to the posterior surface of the upper arm and forearm, to the dorsal surface of the thumb, and the skin as far as the middle of the second finger.

In paralysis of this nerve there is inability to extend the relaxed muscles of the hand and fingers, as well as to extend and abduct the thumb. The outstretched arm cannot be supinated (but on flexion of the arm the forearm can be supinated by the biceps). Such a paralysis is observed in lead paralysis, except that the supinator longus is generally exempt. The sensory disturbances in paralysis of the nerves of the arm may be inferred from the above description of the distribution of the sensory branches, but the symptoms are generally less distinctly marked.

3. *The dorsal nerves* supply the skin and muscles of the thorax and abdomen.

4. *The plexus lumbalis* (12th cervical to 1st-4th dorsal nerve) goes to the skin of the lower abdominal region, of the anterior surface of the thigh, and of the inner surface of the leg. The motor branches supply the internal pelvic muscles. The nervus cruralis supplies the musculus quadriceps femoris, sartorius, pectineus ; the nervus obturatorius supplies the musculus obturator, adductor magnus, longus, and brevis, and gracilis.

5. *The plexus sacralis* (5th lumbar to 1st-5th sacral nerve) supplies the bladder, rectum, sexual organs, perineum, and nates with motor and sensory branches.

The nervus ischiadicus [sciatic nerve], which supplies the skin on the posterior surface of the thigh, on the outer side of the leg, and on the foot as well as the musculus biceps, semitendinosus and semimembranosus, divides half way down the thigh into the nervus tibialis and peroneus, the former of which supplies the muscles on the posterior surface of the leg (calf muscles) and of the under surface of the foot, the latter going to the muscles on the anterior surface of the leg and foot (see Figs. 58 and 59).

CHAPTER XIII.

ANALYSIS OF THE PATHOLOGICAL CON-CREMENTS.

Urinary Concrements.—THE concrement should be rubbed to a fine powder, and a part of it heated red hot on a platinum spatula or on a porcelain crucible top. If the concrement be completely destroyed, or if only a small amount of ash remain behind, then it consists of organic substance, *i. e.*, uric acid, urate of ammonia, xanthin, or cystin.

Uric Acid is tested for with the murexide test, by moistening some of the powder on a crucible lid with a drop of nitric acid and slowly evaporating it. If uric acid is present an orange-red mark is left, which turns purple on being moistened with ammonia. Uric acid calculi are generally reddish-yellow and hard.

Ammonia is tested for by dissolving the powder with dilute hydrochloric acid, filtering and making the filtrate alkaline with caustic potash, and heating it in a test tube. A smell of ammonia is developed, and moist red litmus paper, held over the opening of the tube, turns blue from the vapor ; and a glass rod moistened with muriatic acid and held over the opening of the tube causes a vapor of chloride of ammonia. If uric acid and ammonia are detected the calculus contains urate of ammonia. Such stones are generally white and crumbly.

If the murexide test does not succeed, the *xanthin* is

132

tested for by dissolving the powder in dilute nitric acid, and evaporating it slowly on a porcelain crucible top. If a lemon-colored residue is left which is unchanged on moistening it with ammonia, but turns red on adding caustic potash, then what remains is xanthin. Xanthin calculi are generally of a cinnamon-brown color, moderately hard, and take on a waxy lustre on being rubbed.

Cystin is detected by dissolving the sediment with heat and ammonia. After evaporating, the filtrate may be recognized microscopically as regular hexagonal crystals of cystin. Cystin calculi are generally smooth, yellow, and not very hard.

If the concrement is not completely consumed, but made black only, then it consists of inorganic substances, or of compounds of organic acids (uric or oxalic acid), with alkalies or alkaline earths.

A little of the sediment is put into a test tube, and dilute hydrochloric acid is added. If effervescence take place, it is proof of the presence of carbonic acid. If the substance be not completely dissolved on heating, then the residue may consist of uric acid (to be detected by the murexide test). It should then be filtered, the filtrate made alkaline with ammonia, and then made slightly acid with acetic acid. If then a white powdery precipitate, insoluble with heat, be left, it consists of the *oxalate of lime.* It should then be filtered, and oxalate of ammonia added. A white precipitate shows the presence of calcium monoxide. This is heated and filtered, and ammonia added, and if a precipitate (ammonio-phosphate of magnesia) be formed after standing, it shows the presence of *magnesia* and *phosphoric acid.* If no precipitate be formed, the fluid is divided into two parts, and to one part phosphate of sodium and to the

other sulphate of magnesium is added.   The appearance
of a precipitate in the first shows the presence of *magnesia;*
in the second, of *phosphoric acid.*   Phosphoric acid may
be detected in the fluid by adding acetate of uran after
the acetic acid, and a yellow-white precipitate shows the
presence of the phosphate of uran.

Sulphuric acid may be detected by adding chloride of
barium after the muriatic acid, and a white precipitate of
the sulphide of barium results if sulphuric acid be
present.

Calculi of the oxalate of lime are generally very hard,
of a mulberry shape, and are colored dark with the coloring matter of the blood.   They are not dissolved by
acetic acid ; but are dissolved without effervescence by
the mineral acids.

Calculi of the phosphate of lime and of ammonio-
phosphate of magnesia are generally white, soft, and
friable.

Calculi of the carbonate of lime are white, chalky, and
effervesce on adding acids.

The concrements of the intestine (fæcal calculi)
consist partly of organic substances of different kinds,
and partly of inorganic salts, such as the phosphate of
calcium, the ammonio-phosphate of magnesia, the sulphates of the mineral alkalies.   They should be dissolved
in muriatic acid, and examined in the customary manner.

Salivary calculi generally consist of carbonate of
lime.

Gall stones consist principally of cholesterine and
bilirubin, in combination with calcium.   To detect
cholesterine the powdered concrement should be dissolved in hot alcohol, and filtered, and after cooling the
cholesterine crystallizes out of the filtrate in slender

plates. If the cholesterine be then dissolved in chloroform and concentrated sulphuric acid be added, a beau-
tiful cherry-red color is formed, which changes later to
blue and green. To test for *bilirubin* the residue of the
concrement is made slightly acid with muriatic acid and
extracted with chloroform in a warm place. On adding
fuming nitric acid to the chloroform the Gmelin reaction
appears.

CHAPTER XIV.

METABOLISM AND NUTRITION.

IN order that the human organism shall retain its proper amount of albumen, fat, ashy residue, and water, there must be a sufficient amount and proper proportion of these substances in the daily food. Since the water and ashy residue are generally abundant, the principal question is of giving those nutritious substances which prevent the body from losing albumen and fat. These substances are the albuminous and fatty matters and the carbo-hydrates.

In order to see whether an organism keeps up the proper amount of nourishment or not, it is not sufficient to consider the weight alone ; for the weight can remain the same, or even increase, while the nourishment decreases, as in cases of hydræmia, even without visible œdema.

Albumen is also set free[1] from the organism in hunger. In order that the body shall not lose its proper amount of albumen, a certain quantity should always be given with the food, which can be substituted by no other food. The smallest amount of albumen with which the body can be kept up to its standard is called the diet of sustenance. This for a medium-sized adult is about 85 grams

---

[1] In long-continued complete inanition, about 4.26 grams [63 grains] of nitrogen, equivalent to 210 grams [6¾ ounces] of muscle, are split up and set free daily.

[2½ ounces] (Voit [1]). If more albumen be given more is decomposed, and the body soon sets it free and quickly regains its nitrogenous equilibrium, *i. e.*, just as much is excreted as is taken up. The body possesses extensive powers of adaptation and can reach the nitrogenous equilibrium with the most different amounts of albumen, in case it does not fall below the diet of sustenance. Besides the *amount of albumen necessary for the diet*, the amount of albuminous decomposition is also dependent upon the *body's richness in albumen*, and therefore a muscular working man needs larger amounts of nutritive albumen, than a reduced sick man. The work itself has therefore no influence upon the amount of albumen converted ; for the workman sets free just as much albumen when resting as when working. If less albumen be given with the food than is necessary to keep the normal amount in the body, the organism loses its albumen and becomes poorer in albumen. An increase of albumen in the body cannot be brought about by administering albuminous food only, but in addition there should be large quantities of fat and the carbo-hydrates in the diet, which prevent the albumen from splitting up and thus economize, as it were, the albumen. In *fever* more albumen is decomposed than normally, and the body may therefore lose an enormous amount of albumen, when very little is taken in with the food.

Since the *nitrogen* which arises from the split-up albumen is almost exclusively excreted through the urine (generally as urea, *cf.* p. 70), the amount of albumen converted in the organism may be calculated from the amount of nitrogen in the urine. One gram [15 grains] of nitro-

---

[1] Other authors give lower figures : 56 grams [1¾ ounces], Meinert : 57 grams [1¾ ounces], Playfair.

gen in the urine is equivalent to a change of 6.25 grams [93 grains] of albumen, or of 29.4 grams [449 grains] of muscle (1 gram [15 grains] of urea is equivalent to 2.9 grams [43 grains] of albumen and 13.72 grams [205 grains] of muscle). If the amount of albumen in the food be known and also the amount passed with the fæces, by comparing these figures with the nitrogen excreted in the urine, it may be decided whether the organism has the proper amount of nitrogen, or whether it has gained or lost albumen. If, for example, a patient with typhoid fever take in twenty-four hours 5.977 grams [90 grains] of nitrogen, of which 1.087 grams [16 grains] of nitrogen are passed out with the fæces and 19.488 grams [292 grains] with the urine, the body loses in this time 14.598 grams [218 grains] more than it has assimilated, *i. e.*, it has lost 91.236 grams [1368. grains] of albumen (14.598 × 6.25 grams [218 × 93 grains]) or 429.2 grams [6478. grains] of muscle (14.598 × 29.4 grams [218 × 441 grains]).

A comparison of the amount of urea in disease with that in healthy individuals who are under entirely different conditions of nourishment, is not possible according to the above.

In diseases of the urine-producing organs, as in nephritis, all the products formed in the body from the splitting up of albumen are not always excreted by the kidney, but they may be retained in the body. The danger of poisoning with these matters (uræmia) is greater, as is easily understood, the more albumen taken in as food, and consequently with it the formation of products of albuminous decomposition.

The change of the *non-nitrogenous* substances, *i. e.*, of the fats and carbo-hydrates, is, as opposed to that of the albuminous formations, dependent upon the *amount of work* and the *production of heat.* For example, a laborer, when working, sets free twice as much as when he is resting. In fever when there is an increased production of heat, there is also an increased metabolism of the non-nitrogenous substances. The products of the metabolism of these substances are oxidized to water and carbonic acid and leave the body through the lungs. If less non-nitrogenous food is taken in than the body needs for work and the production of heat, then a part of the fat of the body is used. And if more fat or carbo-hydrates be taken up than are used, then fat is deposited. The fats and carbo-hydrates are the principal non-nitrogenous substances which are taken in as food, and they may take each other's place as food, just in proportion to the amount of heat (calories) which they contribute to combustion. When a certain quantity of albuminous substances (sustenance) is given, never mind what substances they are; 100 grams [1500. grains] of fat have the same value (isodynamic) as 211 grams [3155. grains] of albumen, or 232 grams [3480. grains] of starch, or 234 grams [3510 grains] of cane sugar, or 256 grams [3840 grains] of grape sugar (Rubner).

If the object be to increase the amount of fat in an individual, then a smaller amount of albuminous substances, with a very abundant non-nitrogenous diet should be given, and since fat cannot easily be taken in larger amounts than of 100 grams [1500. grains], the carbo-hydrates should be used. On the contrary, if the wish be to decrease the amount of fat in a body, then the smallest possible quantities of fat and carbo-hydrates with plenty of albuminous food should be given, and care should be taken that the fat be burned up by sufficient exercise (work). In diabetes the organism has lost in a varying degree the ability to split up and use the carbo-hydrates. They are excreted in the urine unused as grape sugar. On the contrary, the body splits up large quantities of albumen and fat, and the loss of the carbo-hydrates therefore must be made up by a diet very rich in albumen and fat, but poor in the carbo-hydrates.

The need of these substances is, therefore, very different for different persons. According to C. v. Voit the amount of food necessary for

	Albumen (grams).	Fat (grams).	Carbo-hydrates (grams).	N. (grams).	C. (grams).
A muscular work-man (of 70 kilo [1] [140 lbs.])	118 [3¾ oz.]	56 [1¾ oz.]	500 [15¼ oz.]	19 [½ oz.]	328 [10 oz.]
A moderately strong man (physician)	127 [4 oz.]	89 [2¾ oz.]	362 [11 oz.]		
An idle man (priso-ner)	87 [2¾ oz.]	22 [¾ oz.]	305 [9 oz.]		

The need of food for growing individuals (children) is less than for adults, but greater in proportion to the weight of the body.

The composition of the most important kinds of food is given in the table opposite.

The nutritive substances are not completely absorbed in the intestinal canal, but a part of them is always passed out unused with the fæces. Under normal conditions animal albumen (meat, eggs, cheese, etc.) is very thoroughly used up, while vegetable albumen (bread, legumes, vegetables) are less thoroughly absorbed. The

[1] According to Pflüger and Bohland the amount of albumen converted is in adults (of 62 kilo [124 lbs.]) equivalent to 96.467 grams [1447. grains].

COMPOSITION OF FOODS.

Food	Dry substance %	Albumen %	N. %	Fat %	Carbo-hydrates %
Raw beef, lean, freed from all visible fat	24.1	18.36	3.4	0.9	[1]
Raw beef, moderately fat	27.75	20.91		5.19	0.48 [2]
Raw beef, fat	44.58	17.19		26.38	[2]
Boiled beef	24.2	21.8		0.9	[3]
Roast beef	41.43		4.89	6.78	[4]
Raw veal	27.69	18.88		7.41	0.07 [2]
Roast veal	21.0	15.3		5.2	[3]
Raw chicken (lean)	23.78	19.72		1.4	1.27 [2]
Hens' eggs (after taking off the shell)	26.1	14.1	2.19	10.9	[1]
1 egg weighs, without the shell, on an average, 45 grams [1⅓ ounce].					
Cow's milk (good quality)	12.92	4.13	0.64	3.90	4.20 [1]
Cow's milk (inferior quality)	11.7	3.5	0.5	2.7	4.5 [5]
Butter	88.3	0.5		87.0	0.5 [2]
Cheese	66.8	32.2	4.75	26.6	2.97 [3]
Bacon				95.6	[6]
Wheat flour (fine)	85.14	8.91	1.42 [4]	1.11	74.28 [2]
White bread	72.0	9.6	1.5 [5]	1.0	60.0 [3]
Black bread	63.29	8.5	1.0		52.5 [1]
Pastry food (the average of seven varieties)	44.2	8.7		15.0	28.9 [3]
Raw potatoes (without skin)	26.62	2.31	0.37		23.3 [4]
Cooked potatoes (without skin)	25.4	2.18	0.35		23.0 [4]
Uncooked peas	86.59	21.25	3.40	1.17	61.8 [4]
Rice	86.5	8.31	1.33		89.2 [4]
Vegetables (average)	27.7	2.2	0.35	3.9	18.1 [5]
Bouillon	0.09		0.057 [5]	0.8	
Soup (the average of ten varieties)	8.4	1.1		1.5	5.7 [5]
White wine	14.0		0.033		2.0
Red wine (French)	11.7		0.0182 [5]		2.34
Sherry	20.5	0.20			3.27 [2]
Bavarian beer	9.7	0.44			5.78 [2]

[1] Analysis of C. v. Voit ; [2] of König ; [3] of Renk ; [4] of Rubner ; [5] of F. Müller ; [6] of Hoffmann.

carbo-hydrates (starch, sugar) are generally very completely used up, while a very large part of the fats is passed out unused with the fæces. In many pathological conditions the absorption of the nutritious substances is badly carried out, as in diarrhœa. In absence of gall in the intestine (icterus), the fats are not easily absorbed.

The absorption in health of some of the most important articles of food is shown in the following table (Rubner) :

Food.	Amount passed out in the fæces of			
	Dry substance (%).	Albumen (N.) (%).	Fat (%).	Carbohydrates (%).
Roast beef	5.06	2.65	19.2	
Eggs	5.2	2.9	5.0	
Milk	9.1	5.9	5.7	0
White bread . . .	4.4	20.7		1.1
Black bread . . .	15.0	32.0		10.9
Pastry	4.9	20.5		1.6
Rice	4.1	20.4		0.9
Potatoes	9.4	32.2		7.6
Vegetables . . .	14.9	18.5	6.1	15.4
Peas	11.5	22.6		5.3
Fat (100 grams [3 ounces] of bacon) . . .	5.5	12.1	17.4	1.6

Finally, may be added a list of important factors necessary for calculation in the experiment of metabolism :

Nitrogen : Urea $= 1 : 2.143$.

Nitrogen : Albumen $= 1 : 6.25$.

Nitrogen : Muscular substance $= 1 : 29.4$.

Urea : Nitrogen $= 1 : 0.466$.

Urea : Albumen $= 1 : 2.9$.

Urea : Muscular substance $= 1 : 13.71$.

Muscular substance : Nitrogen $= 1 : 0.034$.

Albumen : Nitrogen $= 1 : 0.16$.

TABLE OF THE WEIGHTS OF THE HUMAN BODY.

MALE.

Age.	Height in ft. and inches.	Weight.
At birth.	1 ft. 7 in. [0.496 m.]	7 lbs. [3.20 kg.]
1 year.	2 ft. 3 in. [0.696 m.]	22 " [10.00 "
2 "	2 ft. 8 in. [0.797 m.]	26 " [12.00 "
3 "	2 ft. 9 in. [0.860 m.]	29 " [13.21 "
4 "	3 ft. [0.932 m.]	33 " [15.07 "
5 "	3 ft. 3 in. [0.990 m.]	36 " [16.70 "
6 "	3 ft. 5 in. [1.046 m.]	39 " [18.04 "
7 "	3 ft. 8 in. [1.112 m.]	44 " [20.16 "
8 "	3 ft. 9 in. [1.170 m.]	49 " [22.26 "
9 "	4 ft. [1.227 m.]	53 " [24.09 "
10 "	4 ft. 2 in. [1.282 m.]	57 " [26.12 "
12 "	4 ft. 6 in. [1.359 m.]	68 " [31.00 "
14 "	4 ft. 10 in. [1.487 m.]	89 " [40.50 "
16 "	5 ft. 3 in. [1.610 m.]	117 " [53.39 "
18 "	5 ft. 6 in. [1.700 m.]	135 " [61.26 "
20 "	5 ft. 8 in. [1.711 m.]	143 " [65.00 "
25 "	5 ft. 9 in. [1.722 m.]	150 " [68.29 "
30 "	5 ft. 9 in. [1.722 m.]	152 " [68.90 "
40 "	5 ft. 8 in. [1.713 m.]	151 " [68.81 "
50 "	5 ft. 6 in. [1.674 m.]	148 " [67.45 "
60 "	5 ft. 5 in. [1.664 m.]	144 " [65.50 "]

FEMALE.

Age.	Height in ft. and inches.	Weight.
At birth.	1 ft. 6 in. [0.483 m.]	6 lbs. [2.91 kg.]
1 year.	2 ft. 3 in. [0.690 m.]	20 " [9.30 "
2 "	2 ft. 6 in. [0.780 m.]	25 " [11.40 "
3 "	2 ft. 9 in. [0.850 m.]	27 " [12.45 "
4 "	3 ft. [0.910 m.]	31 " [14.18 "
5 "	3 ft. 2 in. [0.974 m.]	34 " [15.50 "
6 "	3 ft. 4 in. [1.032 m.]	37 " [16.74 "
7 "	3 ft. 7 in. [1.096 m.]	40 " [18.45 "
8 "	3 ft, 9 in. [1.139 m.]	43 " [19.82 "
9 "	3 ft. 11 in. [1.200 m.]	50 " [22.44 "
10 "	4 ft. 1 in. [1.248 m.]	53 " [24.24 "
12 "	4 ft. 4 in. [1.327 m.]	67 " [30.54 "
14 "	4 ft. 9 in. [1.447 m.]	84 " [38.10 "
16 "	4 ft. 11 in. [1.500 m.]	98 " [44.44 "
18 "	5 ft. 1 in. [1.562 m.]	117 " [53.10 "
20 "	5 ft. 2 in. [1.570 m.]	120 " [54.46 "
25 "	5 ft. 2 in. [1.577 m.]	121 " [55.08 "
30 "	5 ft. 2 in. [1.579 m.]	121 " [55.14 "
40 "	5 ft. 1 in. [1.555 m.]	129 " [58.45 "
60 "	5 ft. [1.516 m.]	125 " [56.73 "]

DOSE TABLE.

[The doses given are for adults. For hypodermic use the dose should be one half, and for use by the rectum, twice that by the mouth. The dose for children is calculated by adding 12 to the age of the child, and dividing by the age, thus : for a child 4 years old the dose would be $\frac{12 \times 4}{4} = 4$ or $\frac{1}{4}$ of the dose for adults. The doses are given in terms, both of the Apothecaries' and of the Decimal metric system.

Remedies.	Dose expressed in terms of apothecaries' weights and measures.	Dose expressed in metric terms.
Acet. colchici	f ℥ ss to ℥ i	2 to 4 ccm.
" lobeliæ	♏ xv to lx	1 to 4 ccm.
" opii	♏ v to x	0.30 to 0.60 ccm.
" sanguinar.	♏ xv to xx	1 to 2 ccm.
" scillæ	♏ x to xxx	0.60 to 2 ccm.
Acid. acet. dil.	♏ lx to xc	4 to 6 ccm.
" arsenios.	gr. $\frac{1}{64}$ to $\frac{1}{12}$	0.001 to 0.005 gm.
" benzoic.	gr. v to xv	0.30 to 1 gm.
" boracic.	gr. v to x	0.30 to 0.60 gm.
" carbolic.	gr. j to iij	0.05 to 0.20 gm.
" gallic.	gr. iij to xv	0.20 to 1 gm.
" gall. in albuminuria . .	gr. x to lx	0.60 to 4 gm.
" hydrobrom. (34 ℥) . . .	gr. x to xv	0.60 to 1 gm.
" hydrobrom. dil. . . .	♏ xv to xl	1 to 4 ccm.
" hydrochlor. dil. . . .	♏ x to xxx	0.60 to 2 ccm.
" hydrocyan. dil. . . .	♏ ij to vj	0.10 to 0.40 ccm.
" lactic.	gr. xv to lx	1 to 4 gm.
" nitr. dil. . . .	♏ x to xxx	0.30 to 2 ccm.
" nitro-hydrochlor. dil. . .	♏ v to xx	0.30 to 1.20 ccm.
" phosphoric. (50 ℥) . .	gr. iij to v	0.20 to 0.30 gm.
" phosphoric. dil. . .	♏ x to lx	0.60 to 4 ccm
" salicyl.	gr. v to xv	0.30 to 1 gm.
" sulphuric. dil. . . .	♏ v to xxx	0.30 to 2 ccm.
" sulphuric. arom. . . .	♏ v to xxx	0.30 to 2 ccm.
" sulphuros.	♏ xxx to lx	2 to 4 ccm.
" tannic.	gr. ij to x	0.10 to 0.60 gm.

Remedies.	Dose expressed in terms of apothecaries' weights and measures.	Dose expressed in metric terms.
Aconitina (white crystals) . .	gr. $\frac{1}{400}$ to $\frac{1}{200}$	0.00015 to 0.0005 gm.
Adonidine	gr. $\frac{1}{10}$ to $\frac{1}{3}$	0.06 to 0.02 gm.
Aloe	gr. ii to v	0.10 to 0.30 gm.
Aloe et canella	gr. v to xxx	0.30 to 2.0 gm.
Aloinum	gr. j to iij	0.06 to 0.20 gm.
Alumen (expectorant) . . .	gr. iij to x	0.20 to 0.60 gm.
" exsiccat.	gr. v to xxx	0.30 to 2 gm.
Ammonii benzoas	gr. x to xx	0.60 to 1.2 gm.
" bromid.	gr. v to xxx	0.30 to 2 gm.
" carb.	gr. iij to x	0.20 to 0.60 gm.
" chlorid.	gr. xv to xxx	1 to 20 gm.
" iodid.	gr. iij to xv	0.20 to 1.0 gm.
" phosph.	gr. v to xx	0.30 to 1.2 gm.
" picras	gr. $\frac{1}{4}$ to $\frac{1}{2}$	0.15 to 0.03 gm.
" sulph.	gr. iij to xv	0.20 to 1.0 gm.
" valer.	gr. iij to xv	0.20 to 1.0 gm.
Amyl nitris	♏ ij to v	0.10 to 0.30 gm.
Anthemis	ʒ ss to ʒ j	2.0 to 4.0 gm.
Antimonii et pot. tartr. (diaph.) .	gr. $\frac{1}{18}$ to $\frac{1}{6}$	0.004 to 0.01 gm.
" et pot. tartr. (emetic) .	gr. j to ij	0.05 to 0.10 gm.
" oxid. . . .	gr. j to ij	0.05 to 0.10 gm.
" oxysulphuret. . .	gr. $\frac{1}{2}$ to ii	0.03 to 0.10 gm.
" sulphid. . . .	gr. $\frac{1}{2}$ to ij	0.03 to 0.01 gm.
" sulphuret. . . .	gr. $\frac{1}{2}$ to ii	0.03 to 0.10 gm.
Antipyrine	gr. v to xv	0.30 to 1.0 gm.
Apomorph. hydrochlor. . . .	gr. $\frac{1}{30}$ to $\frac{1}{10}$	0.002 to 0.005 gm.
Aqua ammoniæ	♏ vj to xxx	0.40 to 2 ccm.
" amygd. amar. . . .	f ʒ ij to xv	8.0 to 16.0 ccm.
" camphoræ	f ʒ ss to ij	15 to 60 ccm.
" chlori	f ʒ j to iv	4 to 15 ccm.
" creasoti	f ʒ j to iv	4 to 15 ccm.
" laurocerasi	♏ vj to xxx	0.40 to 2 ccm.
Argenti iodidum	gr. $\frac{1}{2}$ to ij	0.03 to 0.10 gm.
" nitras	gr. $\frac{1}{6}$ to $\frac{1}{3}$	0.01 to 0.02 gm.
" oxid.	gr. $\frac{1}{2}$ to ij	0.03 to 0.10 gm.
Arnica	gr. v to xx	0.30 to 1.20 gm.
Arsenici iodidum	gr. $\frac{1}{64}$ to $\frac{1}{10}$	0.001 to 0.005 gm.
Asafœtida	gr. v to xx	0.30 to 1.20 gm.
Atropina	gr. $\frac{1}{128}$ to $\frac{1}{32}$	0.005 to 0.002 gm.
Atropinæ sulph.	gr. $\frac{1}{128}$ to $\frac{1}{32}$	0.005 to 0.002 gm.

Remedies.	Dose expressed in terms of apothecaries' weights and measures.	Dose expressed in metric terms.
Auri et sodii chlorid. . . .	gr. $\frac{1}{32}$ to $\frac{1}{16}$	0.002 to 0.004 gm.
Belladonnæ folium	gr. j	0.05 gm.
Bismuthi citras	gr. iij to xv	0.20 to 1.0 gm.
" et ammon. citr. . .	gr. j to xv	0.05 to 1.0 gm.
" sub-carb.	gr. vj to xxx	0.40 to 2 gm.
" sub-nitr.	gr. vj to xxx	0.40 to 2 gm.
" tannas	gr. vj to xxx	0.40 to 2 gm.
" valer.	gr. j to iij	0.05 to 0.20 gm.
Brucina	gr. $\frac{1}{64}$ to $\frac{1}{16}$	0.001 to 0.004 gm.
Buchu	gr. xx to xxx	1.20 to 2.0 gm.
Caffeina	gr. j to v	0.05 to 0.30 gm.
Caffeinæ citras	gr. j to v	0.05 to 0.30 gm.
Calcii bromidum	gr. v to xxx	0.30 to 2.0 gm.
" carb.	gr. xv to lx	1 to 4 gm.
" hypophosphis . . .	gr. iij to xv	0.20 to 1 gm.
" iodidum	gr. j to iij	0.05 to 0.20 gm.
" phosphas	gr. xv to xxx	1 to 2 gm.
Calumba	gr. x to xxx	0.60 to 2.0 gm.
Calx sulphurata	gr. $\frac{1}{3}$ to j	0.02 to 0.05 gm.
Cambogium	gr. I to iv	0 05 to 0.25 gm.
Camphora	gr. iij to x	0.20 to 0.60 gm.
Camph. monobrom. . . .	gr. ij to v	0.10 to 0.30 gm.
Cantharis	gr. $\frac{1}{2}$ to ij	0.03 to 0.10 gm.
Cardamonum	gr. v to xxx	0.30 to 2 gm.
Castoreum	gr. vj to xv	0.40 to 1 gm.
Catechu	gr. xv to xxx	1 to 2 gm.
Cerii nitras	gr. j to iij	0.05 to 0.20 gm.
" oxalas	gr. j to iij	0.05 to 0.20 gm.
Chinoidinum	gr. iij to xxx	0.20 to 2 gm.
Chloral hydrat. . . .	gr. iij to xv	0.20 to 1 gm.
Chloroformum	♏ j to v	0.05 to 0.30 ccm.
Cinchona	gr. xv to lx	1 to 4 gm.
Cinchonidina and its salts	gr. j to xxx	0.05. to 2 gm.
Cinchonina and its salts .	gr. j to xxx	0.05 to 2 gm.
Cinnamonum	gr. vj to xxx	0.40 to 2 gm.
Codeina	gr. $\frac{1}{2}$ to ij	0.03 to 0.10 gm.
Colchici radix	gr. ij to vj	0.03 to 0.40 gm.
" semen	gr. ij to vj	0.03 to 0.40 gm.
Colocynthis	gr. v to xv	0.30 to 1.0 gm.
Confectio opii	gr. x to xx	0.60 to 1.20 gm.

Remedies.	Dose expressed in terms of apothecaries' weights and measures.	Dose expressed in metric terms.
Confectio sennæ	gr. j to ij	0.5 to 0.10 gm.
Conii fol.	gr. iij	0.20 gm.
Coniina and its salts . . .	gr. $\frac{1}{64}$ to $\frac{1}{32}$	0.001 to 0.002 gm.
Copaiba	ℳ xv to lx	1 to 4 ccm.
Creasotum	ℳ j to iij	0.05 to 0.20 ccm.
Creta præpar.	gr. xv to lxxv	1 to 5 gm.
Croton chloral	gr. j to x	0.05 to 0.60 gm.
Cubeba	gr. xv. to lx	1 to 4 gm.
Cupri acetas	gr. $\frac{1}{2}$ to vj	0.03 to 0.40 gm.
" sulphas	gr. $\frac{1}{2}$ to x	0.03 to 0.60 gm.
Cuprum ammon.	gr. $\frac{1}{8}$ to j	0.01 to 0.05 gm.
Curare	gr. $\frac{1}{32}$ to $\frac{1}{6}$	0.002 to 0.01 gm.
Decoct. aloes comp. . . .	f ℥ ss to ij	15 to 60 ccm.
" chimaphilæ . . .	f ℥ ij	60 ccm.
" citronæ	f ℥ ij	60 ccm.
" sarsap comp. . . .	f ℥ ij to vj	50 to 200 ccm.
Digitalinum	gr. $\frac{1}{64}$ to $\frac{1}{32}$	0.001 to 0.002 gm.
Digitalis	gr. $\frac{1}{2}$ to ij	0.03 to 0.10 gm.
Duboisina and its salts . .	gr. $\frac{1}{128}$ to $\frac{1}{60}$	0.0005 to 0.001 gm.
Elaterinum, U. S. P., 1880 .	gr. $\frac{1}{60}$ to $\frac{1}{16}$	0.001 to 0.004 gm.
Elaterium " 1870 .	gr. $\frac{1}{16}$ to $\frac{1}{2}$	0.004 to 0.03 gm.
Emetina and salts, (emetic) .	gr. $\frac{1}{8}$ to $\frac{1}{4}$	0.008 to 0.016 gm.
" and salts, (diaph.) .	gr. $\frac{1}{120}$ to $\frac{1}{30}$	0.0005 to 0.002 gm.
Ergota	gr. xv to lx	1 to 4 gm.
Ergotinum	gr. ij to viij	0.10 to 0.50 gm.
Eserinæ and its salts . .	gr. $\frac{1}{64}$ to $\frac{1}{20}$	0.001 to 0.004 gm.
Extr. absinthii fl. . . .	ℳ xv to xxx	1 to 2 ccm.
" aconiti fol. (Engl.) . .	gr. $\frac{1}{3}$ to j	0.02 to 0.05 gm.
" aconiti fol., U. S. P., 1870 .	gr. $\frac{1}{2}$ to ij	0.03 to 0.10 gm.
" aconiti fol. fluid. . .	ℳ i to v	0.05 to 0.30 ccm.
" aconiti rad., U. S. P., 1880 .	gr. $\frac{1}{12}$ to $\frac{1}{4}$	0.005 to 0.015 gm.
" aconiti rad. fluid . .	ℳ $\frac{1}{2}$ to ijss	0.03 to 0.13 ccm.
" agaric fl.	ℳ v to xx	0.30 to 1.20 ccm.
" aloes aquos. . . .	gr. $\frac{1}{2}$ to iij	0.03 to 0.20 gm.
" anthemidis	gr. ij to x	0.10 to 0.60 gm.
" anthemidis fl. . . .	ℳ xxx to lx	2 to 4 ccm.
" arnicæ flor. . . .	gr. iij to viij	0.20 to 0.50 gm.
" arnicæ fl.	ℳ v to xv	0.30 to 1 ccm.
" arnicæ rad. . . .	gr. ij to v	0.10 to 0.30 gm.
" arnicæ rad. fl. . . .	ℳ v to xv	0.30 to 1 ccm.

Remedies.	Dose expressed in terms of apothecaries' weights and measures.	Dose expressed in metric terms.
Extr. aromat. fl.	♏ xxx to lx	2 to 4 ccm.
" aurantii. cort. fl.	f ʒ $\frac{1}{4}$ to ijss	1 to 10 ccm.
" bellad. fol. (Engl.)	gr. $\frac{1}{6}$ to $\frac{2}{3}$	0.01 to 0.04 gm.
" bellad. alcohol	gr. $\frac{1}{6}$ to $\frac{1}{2}$	0.01 to 0.03 gm.
" bellad. fol. fl.	♏ iij to vj	0.20 to 0.40 ccm.
" bellad. rad.	gr. $\frac{1}{8}$ to $\frac{1}{4}$	0.008 to 0.016 gm.
" bellad. rad. fl.	♏ j to iij	0.05 to 0.20 ccm.
" berber. aquifol. fl.	♏ xv to xxx	1 to 2 ccm.
" berber. vulg. fl.	♏ xv to xxx	1 to 2 ccm.
" brayeræ fl.	f ʒ ij to iv	8 to 16 ccm.
" bryoniæ fl.	♏ xv to lx	1 to 4 ccm.
" buchu fl.	f ʒ ss to ijss	2 to 10 ccm.
" calami. fl.	♏ xv to lx	1 to 4 ccm.
" calend. fl.	♏ xv to lx	1 to 4 ccm.
" calumbæ	gr. iij to x	0.20 to 1.20 gm.
" calumbæ fl.	♏ xv to lx	1 to 4 ccm.
" cannab. Amer. fl.	♏ iij to xv	0.20 to 1 ccm.
" cannab. ind.	gr. $\frac{1}{6}$ to $\frac{1}{2}$	0.01 to 0.03 gm.
" cannab. ind. fl.	♏ iij to vj	0.20 to 0.40 ccm.
" cantharidis fl.	♏ j to iij	0.05 to 0.20 ccm.
" capsici fl.	♏ j to iij	0.05 to 0.20 ccm.
" cardam. comp. fl.	♏ xv to xlv	1 to 3 ccm.
" carnis	gr. xv to lx	1 to 4 gm.
" cascarillæ fl.	f ʒ $\frac{3}{4}$ to ijss	3 to 10 ccm.
" castaneæ fl.	f ʒ $\frac{3}{4}$ to ijss	3 to 10 ccm.
" catechu liquid.	♏ viij to xxx	0.50 to 2 ccm.
" caulophylli fl.	♏ xv to xxx	1 to 2 ccm.
" cimicifugæ fl.	♏ viij to xxx	0.50 to 2 ccm.
" cinchonæ	gr. xv to xxx	1 to 2 gm.
" cinchonæ fl.	♏ xxx to lx	2 to 4 ccm.
" cinchonæ arom. fl.	♏ xxx to lx	2 to 4 ccm.
" cinchonæ comp. fl.	f ʒ ss to jss	2 to 6 ccm.
" colch. rad.	gr. $\frac{1}{3}$ to $1\frac{1}{2}$	0.02 to 0.08 gm.
" colch. rad. fl.	♏ iij to xv	0.20 to 1 ccm.
" colch. sem. fl.	♏ $1\frac{1}{2}$ to x	0.08 to 1.20 ccm.
" colocynth	gr. $1\frac{1}{2}$ to v	0.08 to 0.30 gm.
" colocynth comp.	gr. $1\frac{1}{2}$ to v	0.08 to 0.30 gm.
" condurango fl.	♏ viij to xxx	0.50 to 2 ccm.
" conii fol. (Engl.)	gr. j to iv	0.05 to 0.25 gm.
" conii fol. alc., U. S. P., 1870,	gr. 1 to $1\frac{1}{2}$	0.05 to 0.08 gm.

Remedies.	Dose expressed in terms of apothecaries' weights and measures.	Dose expressed in metric terms.
Extr. con. (fr.) alc., U. S. P., 1880 .	gr. ⅓ to j	0.02 to 0.05 gm.
" conii fol. fl.	♏ iij to xv	0.20 to 1 ccm.
" con. (fr.) fl. U. S. P., 1880 .	♏ 1½ to v	0 03 to 0.30 ccm.
" convallariæ rad. fl. . . .	♏ xv to xxx	1 to 2 ccm.
" cubebæ fl.	♏ xv to xxv	1 to 2 ccm.
" damianæ fl.	f ℥ ss to ijss	2 to 10 ccm.
" delphinii fl.	♏ j to iij	0.05 to 0.20 ccm.
" digitalis	gr. ⅛ to ½	0.01 to 0.03 gm.
" digitalis fl.	♏ 1 to vj	0.05 to 0.40 ccm.
" duboisiæ	gr. ¼ to ½	0.015 to 0.03 gm.
" duboisiæ fl.	♏ v to x	0.30 to 1.20 ccm.
" dulcamaræ	gr. v. to xv	0.30 to 1 gm.
" dulcamaræ fl.	f ℥ j to ij	4 to 8 ccm.
" ergotæ	gr. iss to viij	0.08 to 0.5 gm.
" ergotæ fl.	♏ xv to lx	1 to 4 ccm.
" erythroxyli fl. . . .	f ℥ ss to ij	2 to 8 ccm.
" eucalypti fl.	♏ xv to lx	1 to 4 ccm.
" euonymi fl.	♏ xv to lx	1 to 4 ccm.
" eupatorii fl.	♏ xxx to lx	2 to 4 ccm.
" euphorb. ipec. fl. . . .	♏ v to xxx	0.30 to 2 ccm.
" ferri. pom.	gr. iij to xv	0.20 to 1 gm.
" frangulæ fl.	f ℥ ss to ijss	2 to 10 ccm.
" fuci vesiculos. . . .	♏ xv to xxx	1 to 2 ccm.
" gallæ fl.	f ℥ ss to ij	2 to 8 ccm.
" gelsemii	♏ ij to viij	0.10 to 0.50 ccm.
" gelsemii fl.	♏ j to viij	0.05 to 0.50 ccm.
" gent. fl.	♏ xxx to lx	2 to 4 ccm.
" gent. comp. fl. . . .	♏ xxx to lx	2 to 4 ccm.
" geranii fl.	♏ xv too xxx	1 to 2 ccm.
" gossypii fl.	♏ xv to xlv	1 to 3 ccm.
" granati. rad. cort. fl. . .	f ℥ ss to ij	2 to 8 ccm.
" grind. rob. fl. . . .	♏ xxx to lx	2 to 4 ccm.
" guaiaci ligni fl. . . .	♏ xxx to lx	2 to 4 ccm.
" guaranæ fl.	♏ xv to xxx	1 to 2 ccm.
" hæmatoxyli	gr. v to xxx	0.30 to 2 gm.
" hæmatoxyli fl. . . .	♏ xxx to lx	2 to 4 ccm.
" hamamelid fl.	♏ xxx to xc	2 to 6 ccm.
" helleb. nigris	gr. ½ to iij	0.03 to 0.20 gm.
" helleb nigris fl. . . .	♏ v to xv	0.30 to 1 ccm.
" humuli . . , , .	gr. iij to xv	0.20 to 1 gm,

Remedies.	Dose expressed in terms of apothecaries' weights and measures.	Dose expressed in metric terms.
Extr. humuli fl.	♏ iij to xv	0.20 to 1 ccm.
" hydrangeæ fl.	♏ xxx to lx	2 to 4 ccm.
" hydrastis	gr. iij to x	0.20 to 1.20 gm.
" hydrastis fl.	♏ v to xxx	0.30 to 2.0 ccm.
" hyoscyami (Engl.)	gr. j to iv	0.05 to 0.25 gm.
" hyoscyami alc.	gr. j to ij	0.05 to 0.10 gm.
" hyoscyami fol. fl.	♏ iij to xv	0.20 to 1 ccm.
" hyoscyami sem. fl.	♏ ij to viij	0.10 to 0.50 ccm.
" ignatiæ	gr. ⅓ to 1¼	0.02 to 0.65 gm.
" ignatiæ fl.	♏ j to vj	0.05 to 0.40 ccm.
" ipecac fl.	♏ iij to lx	0.20 to 4 ccm.
" iridis versicol.	gr. iij to vj	0.20 to 0.40 gm.
" iridis versicol fl.	♏ xv too xxx	1 to 2 ccm.
" jalapæ, U. S. P., 1870	gr. v to x	0.30 to 0.60 gm.
" jalapæ alc.	gr. iij to vj	0.20 to 0.40 gm.
" jalapæ fl.	♏ xv to lx	1 to 4 ccm.
" junip. fl.	♏ xxx to lx	2 to 4 ccm.
" kamala fl.	♏ xxx to lx	2 to 4 ccm.
" kino liquid	♏ xv to xxx	1 to 2 ccm.
" kramariæ	gr. v to xv	0.30 to 1 gm.
" lactucæ fl.	♏ xv to lx	1 to 4 ccm.
" lactucarii fl.	♏ v to xxx	0.30 to 2.0 ccm.
" leptandræ	gr. iij to x	0.20 to 0.60 gm.
" leptrandræ fl.	♏ xxx to lx	2 to 4 ccm.
" lobeliæ fl.	♏ j to v	0.05 to 0.30 ccm.
" lupulini fl.	♏ v to xv	0.30 to 1 ccm.
" matico fl.	♏ xxx to lx	2 to 4 ccm.
" myricæ fl.	♏ xxx to lx	2 to 4 ccm.
" nectandræ fl.	f ℥ j to iv	4 to 16 ccm.
" nuc. vom.	gr. ⅓ to 1½	0.02 to 0.08 gm.
" nuc. vom. fl.	♏ 1 to iv	0.05 to 0.30 ccm.
" opii	gr. ⅙ to ½	0.01 to 0.03 gm.
" papaveris	gr. ½ to ij	0.03 to 0.10 gm.
" papaveris fl.	♏ xv to xlv	1 to 3 ccm.
" pareiræ fl.	♏ xxx to lx	2 to 4 ccm.
" petroselini fl.	f ℥ j to ij	4 to 8 ccm.
" physostigmæ	gr. 1/16 to ⅙	0.004 to 0.01 gm.
" physostigmæ fl.	♏ j to iij	0.05 to 0.20 ccm.
" phytolaccæ baccar. fl.	♏ v to xxx	0.30 to 2 ccm.
" phytolaccæ rad.	gr. j to iij	0.05 to 0.20 gm.

Remedies.	Dose expressed in terms of apothecaries' weights and measures.	Dose expressed in metric terms.
Extr. phytolaccæ rad. fl. . .	ℳv to xxx	0.30 to 2 ccm.
" pilocarpi fl.	ℳxv to lx	1 to 4 ccm.
" pimentæ fl.	ℳxv to xlv	1 to 3 ccm.
" piper nigr. fl.	ℳxv to xlv	1 to 3 ccm.
" podophylli	gr. ¼ to 1¼	0.03 to 0.08 gm.
" podophylli fl.	ℳv to xxx	0.30 to 2.0 ccm.
" polygoni fl.	ℳxv to xxx	1 to 2 ccm.
" polygonati fl.	ℳv to xv	0.30 to 1 ccm.
" prun. virg. fl.	ℳxxx to lx	2 to 4 ccm.
" quassiæ	gr. j to v	0.05 to 0.30 gm.
" quassiæ fl.	ℳxxx to lx	2 to 4 ccm.
" quebracho fl.	ℳx to lx	0.60 to 4 ccm.
" quercus fl.	ℳxxx to lx	2 to 4 ccm.
" rhamni cath. ft. fl. . .	ℳxxx to lx	2 to 4 ccm.
" rhamni pursh. cort. fl. . .	ℳxxx to cxx	2 to 8 ccm.
" rhei	gr. v to xv	0.30 to 1 gm.
" rhei fl.	ℳxv to xlv	1 to 3 ccm.
" ricini fol. fl.	f ℥ ss to ij	2 to 8 ccm.
" rutæ fl.	ℳxv to xxx	1 to 2 ccm.
" sabinæ fl.	ℳv to xv	0.30 to 1 ccm.
" sanguin. fl.	ℳv to xv	0.30 to 1 ccm.
" santali citr. fl. . . .	f ℥ j to ij	4 to 8 ccm.
" santonicæ fl.	ℳxv to lx	1 to 4 ccm.
" sarsap. fl.	f ℥ ss to ij	2 to 8 ccm.
" sarsap. comp. fl. . . .	f ℥ ss to ij	2 to 8 ccm.
" sassafras fl.	f ℥ ss to ij	2 to 8 ccm.
" scillæ fl.	ℳv to xxx	0.30 to 2 ccm.
" scillæ comp. fl. . . .	ℳv to xxx	0.30 to 2 ccm.
" scoparii fl.	f ℥ ss to j	2 to 4 ccm.
" senegæ fl.	ℳv to xv	0.30 to 1 ccm.
" sennæ fl.	f ℥ j to iv	4 to 16 ccm.
" serpent. fl.	ℳxxx to lx	2 to 4 ccm.
" simarubæ fl.	ℳxv to xxx	1 to 2 ccm.
" spigeliæ fl.	ℳxv to lx	1 to 4 ccm.
" spigeliæ et sennæ fl. . .	f ℥ ss to ij	2 to 8 ccm.
" stillingiæ fl.	f ℥ ss to ij	2 to 8 ccm.
" stillingiæ comp. . . .	f ℥ ss to ij	2 to 8 ccm.
" stramonii (Engl.) . . .	gr. ½ to j	0.03 to 0.05 gm.
" stramonii fol. alc. . . .	gr. ⅓ to ⅔	0.02 to 0.04 gm.
" stramonii sem. , , .	gr. ⅙ to ½	0.01 to 0.03 gm.

Remedies.	Dose expressed in terms of apothecaries' weights and measures.	Dose expressed in metric terms.
Extr. stramonii fl.	℔j to vi	0.05 to 0.40 ccm.
" sumbul fl.	℔xv to lx	1 to 4 ccm.
" taraxaci	gr. v to xv	0.30 to 1 gm.
" taraxaci fl.	f ʒ ss to ij	2 to 8 ccm.
" toxicodendri fl. . . .	℔j to v	0.05 to 0.30 ccm.
" trifol. prat. fl. . . .	f ʒ j to ij	4 to 8 ccm.
" urticæ rad. fl. . . .	℔v to xv	0.30 to 1 ccm.
" ustilag. maid. fl. . . .	℔xv to lx	1 to 4 ccm.
" uvæ ursi fl.	℔xxx to lx	2 to 4 ccm.
" valer.	gr. v to xv	0.30 to 1 gm.
" valer. fl.	℔xxx to lx	2 to 4 ccm.
" veratr. vir. fl. . . .	℔ij to viij	0.10 to 0.50 ccm.
" verbenæ fl.	℔xv to lx	1 to 4 ccm.
" viburni opuli fl. . . .	f ʒ j to ij	4 to 8 ccm.
" viburni [prunifol] fl. . .	f ʒ j to ij	4 to 8 ccm.
" yerbæ santæ fl. . . .	f ʒ ¼ to ʒ j	1 to 4 ccm.
" zingiberis fl.	℔v to xxx	0.30 to 2 ccm.
Ferri arsen.	gr. 1/20 to ½	0.003 to 0.03 gm.
" benzoas	gr. j to v	0.05 to 0.30 gm.
" bromid.	gr. j to v	0.05 to 0.30 gm.
" carb. sacch.	gr. iv to xv	0.25 to 1 gm.
" chlorid.	gr. j to iij	0.05 to 0.20 gm.
" citr.	gr. v to x	0.30 to 0 60 gm.
" et ammon. citr. . . .	gr. v to x	0.30 to 0.60 gm.
" et ammon. sulph. . .	gr. v to x	0.30 to 0.60 gm.
" et ammon. tartr. . . .	gr. v to xv	0.30 to 1 gm.
" et cinchonid. citr. . .	gr. v to x	0.30 to 0.60 gm.
" et pot. tartr. . . .	gr. xv to lx	1 to 4 gm.
" et quin. citr. . . .	gr. v to x	0.30 to 0.60 gm.
" et strychnin. citr. . .	gr. j to xv	0.05 to 1 gm.
" ferrocyanid. . . .	gr. iij to v	0.20 to 0 30 gm.
" hypophosphis	gr. v to x	0.30 to 0 60 gm.
" iodidum	gr. j to v	0.05 to 0.30 gm.
" iodidum sacch. . . .	gr. ij to x	0.10 to 0.60 gm.
" lactas	gr. j to iij	0.05 to 0.20 gm.
" oxalas	gr. j to iij	0.05 to 0.20 gm.
" oxid. hydrat. . . .	ʒ ss to ij	15 to 60 ccm.
" oxid. magnet. . . .	gr. v to x	0.30 to 0.60 gm.
" phosphas	gr. j to v	0.05 to 0.30 gm.
" hypophosphas . . .	gr. j to v	0.05 to 0.30 gm.

Remedies.	Dose expressed in terms of apothecaries' weights and measures.	Dose expressed in metric terms.
Ferri sub-carb.	gr. v to xxx	0.30 to 2 gm.
" sulphas	gr. j to iij	0.05 to 0.20 gm.
" sulphas exsiccat. . . .	gr. ½ to 1½	0.03 to 0.08 gm.
" valer.	gr. j to iij	0.05 to 0.20 gm.
Ferrum ammoniat.	gr. v to x	0.30 to 0.60 gm.
" dialys.	ℳj to xv	0.05 to 1 ccm.
" redact.	gr. j to v	0.05 to 0.30 gm.
Filix mas	℈ j to ij	4 to 8 gm.
Fuchsine	gr. j to iij	0.05 to 0.20 gm.
Galla	gr. x to xx	0.60 to 1.20 gm.
Gambogia	gr. ij to iij	0.10 to 0.20 gm.
Gentiana	gr. x to xxx	0.60 to 2 gm.
Guarana	gr. v to xxx	0.30 to 2 gm.
Hydrarg. c. creta	gr. v to x	0.30 to 0.60 gm.
" chlor. corros. . . .	gr. $\frac{1}{64}$ to $\frac{1}{10}$	0.001 to 0.005 gm.
" chlorid. mite . . .	gr. $\frac{1}{6}$ to viij	0.01 to 0.50 gm.
" cyanid.	gr. $\frac{1}{16}$ to $\frac{1}{2}$	0.004 to 0.03 gm.
" iodid. flav. . . .	gr. $\frac{1}{8}$ to j	0.01 to 0.05 gm.
" iodid. rubr. . . .	gr. $\frac{1}{16}$ to $\frac{1}{2}$	0.004 to 0.03 gm.
" iodid. vir. . . .	gr. $\frac{1}{6}$ to j	0.01 to 0.05 gm.
" oxid. flav. . . .	gr. $\frac{1}{16}$ to $\frac{1}{2}$	0.004 to 0.03 gm.
" oxid. nigr. . . .	gr. $\frac{1}{10}$ to j	0.005 to 0.05 gm.
" oxid. rubr. . . .	gr. $\frac{1}{16}$ to $\frac{1}{2}$	0.004 to 0.03 gm.
" subsulphas flav. . .	gr. $\frac{1}{4}$ to j	0.015 to 0.05 gm.
" sulphuret. nigr. . .	gr. v to x	0.30 to 0.60 gm.
" sulphuret. rub. . .	gr. v to x	0.30 to 0.60 gm.
" c. magn.	gr. v to x	0.30 to 0.60 gm.
Infusum brayeræ	f ℥ ij to viij	60 to 250 ccm.
" buchu	f ℥ ij	60 ccm.
" digitalis	f ℥ ij to iv	8 to 16 ccm.
" eupatorii	f ℥ ij	60 ccm.
" sennæ comp. . . .	f ℥ j to ij	30 to 60 ccm.
" ulmi	*Ad libitum.*	*Ad libitum.*
Iodinum	gr. $\frac{1}{4}$ to j	0.015 to 0.05 gm.
Iodoformum	gr. j to iij	0.05 to 0.20 gm.
Ipecacuanha expect. . . .	gr. $\frac{1}{6}$ to j	0.01 to 0.05 gm.
" emet.	gr. xv to xxx	1 to 2 gm.
Jalapa	gr. xv to xxx	1 to 2 gm.
Juniperi baccæ	℈ j to ij	4 to 8 gm.
Kairine	gr. ij to x	0.10 to 0.60 gm.

Remedies.	Dose expressed in terms of apothecaries' weights and measures.	Dose expressed in metric terms.
Kino	gr. x to xxx	0.60 to 2 gm.
Krameria	gr. x to xxx	0.60 to 2 gm.
Lacto-peptine	gr. x	0.60 gm.
Lactucarium	gr. iij to x	0.20 to 0.60 gm.
Liq. ammon. acet.	f ℥ ij to viij	8 to 25 ccm.
" acidi arseniosi	℥ij to vij	0.10 to 0.50 ccm.
" arsen. et hydr. iod. (Donovan's sol.)	℥ij to vij	0.10 to 0.50 ccm.
" ferri chloridi	℥ij to x	0.10 to 0.60 ccm.
" ferri dialys.	℥j to xv	0.05 to 1 ccm.
" ferri nitrat.	℥v to xv	0.30 to 1 ccm.
" pepsini	f ℥ ij to iv	8 to 16 ccm.
" potassæ	℥v to xxx	0.30 to 2 ccm.
" potassii arsenit. (Fowler's solution)	℥iij to vij	0.20 to 0.50 ccm.
" potassii citrat.	f ℥ ij to iv	8 to 16 ccm.
" sodæ	℥v to xxx	0.30 to 2 ccm.
" sodii arseniatis (Pearson's solution)	℥iij to vij	0.20 to 0.50 ccm.
Lithii benzoas	gr. ij to v	0.10 to 0.30 gm.
" bromid.	gr. i to iij	0.05 to 0.20 gm.
" carb.	gr. ij to vi	0.10 to 0.40 gm.
" citr.	gr. ij to v	0.10 to 0.30 gm.
" salicylas	gr. ij to viij	0.10 to 0.50 gm.
Lobelia	gr. v to x	0.30 to 0.60 gm.
Lupulinum	gr. v to x	0.30 to 0.60 gm.
Magnesia	gr. xv to lx	1 to 4 gm.
Magnesii carb.	gr. xv to lx	1 to 4 gm.
" citr. gran. . . .	℥ j to viij	4 to 32 gm.
" sulphas . . .	℥ j to viij	4 to 32 gm.
" sulphis	gr. v to xxx	0.30 to 2 gm.
Manganesii oxid. nigr. (binoxid.) .	gr. ij to x	0.10 to 0.60 gm.
" sulphas	gr. ij to x	0.10 to 0.60 gm.
Manna	℥ i to ij	30 to 60 gm.
Massa copaibæ	gr. v to xxx	0.30 to 2 gm.
" ferri carb.	gr. v to xv	0.30 to 1 gm.
" hydrarg.	gr. i to xv	0.05 to 1 gm.
Mist ammoniaci	f ℥ iv to viij	15 to 30 ccm.
" asafœtidæ	f ℥ iv to viij	15 to 30 ccm.
" chloroformi	f ℥ iv to viij	15 to 30 ccm.
" cretæ	f ℥ j to ij	30 to 60 ccm.
" ferri comp.	f ℥ ss to ij	15 to 60 ccm.
" ferri et ammon. acet. . .	f ℥ ss to j	15 to 30 ccm.

Remedies.	Dose expressed in terms of apothecaries' weights and measures.	Dose expressed in metric terms.
Mist glycyrrh. comp.	f ʒ j to iv	4 to 16 ccm.
" magnes. et asafœt.	f ʒ j to iv	4 to 16 ccm.
" potassii citr.	f ℥ ss to ij	15 to 60 ccm.
" rhei et sodæ	f ℥ ss to j	15 to 30 ccm.
Morphiæ murat.	gr. $\frac{1}{6}$ to $\frac{1}{2}$	0.01 to 0.03 gm.
" sulph.	gr. $\frac{1}{8}$ to $\frac{1}{2}$	0.008 to 0.03 gm.
" acetat.	gr. $\frac{1}{6}$ to $\frac{1}{2}$	0.01 to 0.03 gm.
" sulph. liq.	f ʒ j to iv	4 to 16 ccm.
" sulph. liq. (Magendie)	♏ij to xv	0.10 to 1 ccm.
Moschus	gr. v to x	0.30 to 0.60 gm.
Myrrha	gr. x to xx	0.60 to 1.20 gm.
Napthalin	gr. j to ij	0.05 to 0.10 gm.
Narceina	gr. $\frac{1}{6}$ to ij	0.01 to 0.10 gm.
Nicotia	gr. $\frac{1}{60}$ to $\frac{1}{40}$	0.001 to 0.025 gm.
Nitro-glycerinum	gr. $\frac{1}{64}$ to $\frac{1}{16}$	0.001 to 0.004 gm.
Nux vomica	gr. j to v	0.05 to 0.30 gm.
Oleoresina asphidii	gr xv to lx	1 to 4 gm.
" capsici	gr. $\frac{1}{6}$ to $\frac{1}{2}$	0.01 to 0.03 gm.
" cubebæ	gr. v to xxx	0.30 to 2 gm.
" lupulini	gr. ij to v	0.10 to 0.30 gm.
" piperis	gr. i to iij	0.05 to 0.20 gm.
" zingiberis	gr. j to iij	0.05 to 0.20 gm.
Oleum amygdal amar.	♏ $\frac{1}{8}$ to $\frac{1}{4}$	0.008 to 0.015 ccm.
" anisi	♏ij to v	0.10 to 0.30 ccm.
" cajuput	♏ij to v	0.10 to 0.30 ccm.
" chenopodii	♏v to x	0.30 to 0.60 ccm.
" copaibæ	♏viij to xv	0.50 to 1 ccm.
" cubebae	♏xv to xxx	1 to 2 ccm.
" eriger	♏v to xv	0.30 to 1 ccm.
" eucalypti	♏x to xxx	0.60 to 2 ccm.
" limon.	♏ij to iv	0.10 to 0.20 ccm.
" morrhuæ	f ʒ j to iv	4 to 16 ccm.
" olivæ	f ʒ j to iv	4 to 16 ccm.
" phosphoratum	gr. j to iij	0.05 to 0.20 gm.
" ricini	f ℥ j to iv	4 to 32 ccm.
" sabinæ	♏j to iij	0.05 to 0.20 ccm.
" terebinth.	♏ v to xxx	0.30 to 2 ccm.
" tiglii	♏ $\frac{1}{5}$ to $\frac{1}{2}$	0.01 to 0.08 ccm.
Opium, 14 % morphine	gr. $\frac{1}{6}$ to $1\frac{1}{2}$	0.01 to 0.08 gm.
Pareira	ʒ ss to j	2 to 4 gm.

Remedies.	Dose expressed in terms of apothecaries' weights and measures.	Dose expressed in metric terms.
Paraldehyd.	♏xx to xl	1.20 to 2.40 ccm.
Pelletierine	gr. v to xv	0.30 to 1 gm.
Pepsinum purum	gr. xv to ℥ss	1 to 15 gm.
" saccharatum	gr. xxx to ℥j	2 to 30 gm.
Petroleum	℥ss to j	2 to 4 gm.
Phosphorus	gr. $\frac{1}{125}$ to $\frac{1}{20}$	0.0005 to 0.003 gm.
Physostigminæ salicyl	gr. $\frac{1}{120}$ to $\frac{1}{20}$	0.0005 to 0.003 gm.
" sulphas	gr. $\frac{1}{125}$ to $\frac{1}{20}$	0.0005 to 0.003 gm.
Picrotoxinum	gr. $\frac{1}{64}$ to $\frac{1}{3}$	0.001 to 0.02 gm.
Pilocarpina and salts	gr. $\frac{1}{64}$ to $\frac{1}{2}$	0.001 to 0.03 gm.
Pil. aloes	pil. j to iij	pil. j to iv
" aloes et asafœt.	" ij to v	" ij to v
" aloes et ferri	" j to iij	" j to iij
" aloes et mast.	" j to iij	" j to iij
" aloes et myrrhæ	" ij to v	" ij to v
" antim. comp.	" j to iij	" j to iij
" asafœtidæ	" j to vj	" j to vj
" cathart. comp.	" j to iv	" j to iv
" ferri comp.	" ij to v	" ij to v
" ferri iodidi	" j to iv	" j to iv
" galbani comp.	" j to v	" j to v
" hydrarg.	gr. ss to xv	0.025 to 1 gm.
" opii	pil. j to ij	pil. j to ij
" phosphori	" j to iv	" j to iv
" rhei	" ij to v	" ij to v
" rhei comp.	" ij to v	" ij to v
Piperinum	" gr. j to viij	0.05 to 0.50 gm.
Plumbi acetas	gr. $\frac{1}{2}$ to iij	0.03 to 0.20 gm.
" iodidum	gr. $\frac{1}{4}$ to iij	0.03 to 0.20 gm.
Potassii acetas	gr. xv to lx	1 to 4 gm.
" bicarb.	gr. v to lx	0.30 to 4 gm.
" bichromat.	gr. $\frac{1}{6}$ to $\frac{1}{2}$	0.01 to 0.25 gm.
" bitartr.	gr. j to ij.	0.05 to 0.40 gm.
" bromid.	gr. v to lx	0.30 to 4 gm.
" carb.	gr. v to xxx	0.30 to 2 gm.
" chloras	gr. v to xxx	0.30 to 2 gm.
" citras	gr. xv to lx	1 to 4 gm.
" cyanid.	gr. $\frac{1}{16}$ to $\frac{1}{8}$	0.004 to 0.008 gm
" et sodii tartr.	℥ss to j	15 to 30 gm.
" ferrocyanid.	gr. x to xv	0.60 to 1 gm.

Remedies.	Dose expressed in terms of apothecaries' weights and measures.	Dose expressed in metric terms.
Potassii hypophosphis . . .	gr. v to xv	0.30 to 1 gm.
" iodid.	gr. ij to xv	0.10 to 1 gm.
" nitras	gr. v to xv	0.30 to 1 gm.
" permanganat. . . .	gr. ss to j	0.03 to 0.06 gm.
" sulphas	3 j to iv	4 to 16 gm.
" sulphidum	gr. j to x	0.05 to 0.60 gm.
" sulphis	gr. xv to xxx	1 to 2 gm.
" sulphuret	gr. ij to vj	0.10 to 0.40 gm.
" tartras	3 j to viij	4 to 30 gm.
Prunus virginia	3 ss to j	2 to 4 gm.
Pulv. antimonialis	gr. iij to x	0.20 to 0.60 gm.
" aromat.	gr. v to xxx	0.30 to 2 gm.
" cretæ comp.	gr. v to xxx	0.30 to 2 gm.
" glycyrrh. comp. . . .	gr. xxx to lx	2 to 4 gm.
" ipecac. et opii (Dover) . .	gr. v to xv	0.30 to 1 gm.
" jalapa comp.	gr. xxx to lx	2 to 4 gm.
" morphinæ comp. . . .	gr. v to xv	0.30 to 1 gm.
" rhei comp.	gr. xxx to lx	2 to 4 gm.
Quinidina (and salts) . . .	gr. j to xxx	0.05 to 2 gm.
Quinina (and salts)	gr. j to xxx	0.05 to 2 gm.
Quininæ arsenias	gr. $\frac{1}{6}$ to j	0.01 to 0.05 gm.
Resina copaibæ	gr. ij to x	0.10 to 0.60 gm.
" jalapæ	gr. ij to v	0.10 to 0.30 gm.
" podophylli	gr. $\frac{1}{8}$ to $\frac{1}{2}$	0.008 to 0.03 gm.
" scammonii	gr. ij to x	0.10 to 0.60 gm.
Resorcin	gr. v to xxx	0.30 to 2 gm.
Rheum	gr. ij to xxx	0.10 to 2 gm.
Sabina	gr. v to x	0.30 to 0.60
Salicinum	gr. v to xxx	0.30 to 2 gm.
Santoninum	gr. j to v	0.05 to 0.30 gm.
Sapo	gr. v to xxx	0.30 to 2 gm.
Scammonium	gr. iij to xv	0.20 to 1 gm.
Scilla	gr. i to ij	0.05 to 0.10 gm.
Senega	gr. x to xx	0.60 to 1.20 gm.
Senna	gr. v to lx	0.30 to 4 gm.
Serpentaria	3 j to ij	4 to 8 gm.
Sinapis	3 ij	8 gm.
Sodii acetas	gr. xv to lx	1 to 4 gm.
" arsenias	gr. $\frac{1}{64}$ to $\frac{1}{10}$	0.001 to 0.005 gm.
" benzoas	gr. v to xv	0.30 to 1 gm.

Remedies.	Dose expressed in terms of apothecaries' weights and measures.	Dose expressed in metric terms.
Sodii bicarb.	gr. v to xxx	0.30 to 2 gm.
" bisulphis	gr. v to xxx	0.30 to 2 gm.
" boras	gr. v to xxx	0.30 to 2 gm.
" bromid	gr. v to xxx	0.30 to 2 gm.
" carb.	gr. v to xxx	0.30 to 2 gm.
" carb. exsicc.	gr. v to xv	0.30 to 1 gm.
" chloras	gr. v to xxx	0.30 to 2 gm.
" hypophosphis	gr. v to xv	0.30 to 1 gm.
" hyposulphis	gr. v to xxx	0.30 to 2 gm.
" iodidum	gr. v to xv	0.30 to 1 gm.
" phosphas	gr. ij to xv	0.10 to 1 gm.
" salicylas	gr. v to xxx	0.30 to 2 gm.
" santoninas	gr. ij to x	0.10 to 0.60 gm.
" sulphas	gr. j to ij	0.05 to 0.10 gm.
" sulphis	gr. v. to xxx	0.30 to 2 gm.
Sparteinæ sulph.	gr. ¼ to ½	0.01 to 0.03 gm.
Spigelia	gr. x to ℥ j	0.60 to 30 gm.
Spir. æther. comp. . . .	ℳ xxx to lx	2 to 4 ccm.
" æther. nitrosi . . .	f ℥ ss to ij	2 to 8 ccm.
" ammoniæ	ℳ v to xxx	0.30 to 2 ccm.
" ammoniæ arom. . . .	ℳ xv to xxx	1 to 2 ccm.
" camphoræ	ℳ v to xxx	0.30 to 2 ccm.
" chloroformi	ℳ xv to lx	1 to 4 ccm.
" lavend. comp. . . .	ℳ xxx to lx	2 to 4 ccm.
" menth. pip.	ℳ xxx to lx	2 to 4 ccm.
Stramonii folium	gr. j to ij	0.05 to 0.10 gm.
Strychnina and salts . . .	gr. 1/64 to 1/12	0.001 to 0.005 gm.
Styrax	gr. x to xx	0.60 to 1.20 gm.
Sulphur	℥ ss to iv	2 to 16 gm.
Syr. acidii hydriodidi allii.	f ℥ j to iv	4 to 16 ccm.
" calcii lactophos. . . .	f ℥ j to ij	4 to 8 ccm.
" calcis	ℳ xv to xxx	1 to 2 ccm.
" ferri bromidi . . .	ℳ xv to lx	1 to 4 ccm.
" ferri iodidi . . .	ℳ xv to lx	1 to 4 ccm.
" ferri oxidi	f ℥ j	4 ccm.
" ferri hypophosph. . .	f ℥ j	4 ccm.
" fer. quin. et str. phos. .	f ℥ j	4 ccm.
" hypophosphit. . . .	f ℥ j	4 ccm.
" hypophosph. c. fer. .	f ℥ j	4 ccm.
" ipecac.	f ℥ ss to iv	2 to 16 ccm.

Remedies.	Dose expressed in terms of apothecaries' weights and measures.	Dose expressed in metric terms.
Syr. krameriæ	f ʒ ss to iv	2 to 16 ccm.
" lactucarii	f ʒ j to iij	4 to 12 ccm.
" pruni virginianæ	f ʒ j to ij	4 to 8 ccm.
" rhei	f ʒ j to iv	4 to 16 ccm.
" rhei arom.	f ʒ j to iv	4 to 16 ccm.
" rosæ	f ʒ j to ij	4 to 8 ccm.
" rubi	f ʒ j to ij	4 to 8 ccm.
" sarsap. comp.	f ʒ j to iv	4 to 16 ccm.
" scillæ	f ʒ ss to j	2 to 4 ccm.
" scillæ comp. (hive-sirup)	℥ xv to lx	1 to 4 ccm.
" senegæ	f ʒ j to ij	4 to 8 ccm.
" sennæ	f ʒ j to iv	4 to 16 ccm.
Testa præparata	gr. v to xx	0.30 to 1.20 gm.
Thallin	gr. iv to viij	0.25 to 0.50 gm.
Tinct. aconiti fol.	℥ viij to xvj	0.50 to 1 ccm.
" aconiti rad.	℥ j to v	0.05 to 0 30 ccm.
" acon. rad. (Flemings)	℥ ¾ to ijss	0.04 to 0.15 ccm.
" aloes (1880)	f ʒ ss to ij	2 to 8 ccm.
" aloes et myrrhæ	f ʒ i to ij	4 to 8 ccm.
" arnicæ flor.	℥ v to xxx	0.30 to 2 ccm.
" arnicæ rad.	℥ xv to xxx	1 to 2 ccm.
" asafœtidæ	℥ xxx to lx	2 to 4 ccm.
" belladonnæ	℥ v to xv	0.30 to 1 ccm.
" calumbæ	f ʒ i to iv	4 to 16 ccm.
" cannabis ind.	℥ xv to xxx	1 to 2 ccm.
" cantharid.	℥ v to xv	0.30 to 1 ccm.
" capsici	℥ v to xv	0.30 to 1 ccm.
" cardamomi	f ʒ i	4 ccm.
" cardamomi comp.	f ʒ i	4 ccm.
" catechu comp.	f ʒ ss to ij	2 to 8 ccm.
" cimicifugæ	℥ xxx to lx	2 to 4 ccm.
" cinchonæ	f ʒ ss to ij	2 to 8 ccm.
" cinchonæ comp.	f ʒ ss to ij	2 to 8 ccm.
" colchici rad.	℥ v to xv	0.30 to 1 ccm.
" colchici sem.	℥ v to xv	0.30 to 1 ccm.
" conii	℥ v to xxx	0.30 to 2 ccm.
" cubebæ	f ʒ j to ij	4 to 8 ccm.
" digitalis	℥ v to xv	9.30 to 1 ccm.
" ferri acet.	℥ xv to xxx	1 to 2 ccm.
" ferri chloridi	℥ x to lx	0.60 to 4 ccm.

Remedies.	Dose expressed in terms of apothecaries' weights and measures.	Dose expressed in metric terms.
Tinct. ferri chloridi æther. . .	♏︎xv to xxx	1 to 2 ccm.
" ferri pomati	♏︎xv to lx	1 to 4 ccm.
" gallæ	f ℨ ss to ij	2 to 8 ccm.
" gelsemii	♏︎v to xv	0.30 to 1 ccm.
" gentian comp. . . .	f ℨ ss to ℨ ij	2 to 8 ccm.
" guaiaci	♏︎xxx to lx	2 to 4 ccm.
" guaiaci ammon. . . .	♏︎xxx to lx	2 to 4 ccm.
" hellebori	♏︎x to xv	0.30 to 1 ccm.
" humuli	f ℨ j to ij	4 to 8 ccm.
" hydrastis	♏︎xxx to xc	2 to 6 ccm.
" hyoscyami fol. . . .	♏︎xv to lx	1 to 4 ccm.
" hyoscyami sem. . . .	♏︎xv to xxx	1 to 2 ccm.
" ignatiæ	♏︎v to xv	0.30 to 1 ccm.
" iodini	♏︎v to x	0.30 to 0.60 ccm.
" iodini comp.	♏︎ij to x	0.12 to 0.60 ccm.
" ipecac. et opii . . .	♏︎v to xv	0.30 to 1 ccm.
" jalapæ	f ℨ ss to ij	2 to 8 ccm.
" kino	f ℨ ss to ij	2 to 8 ccm.
" krameriæ	f ℨ ss to ij	2 to 8 ccm.
" lavend. comp. . . .	f ℨ ss to ij	2 to 8 ccm.
" lobeliæ	♏︎xv to xlv	1 to 3 ccm.
" lupulini	f ℨ ss to ij	2 to 8 ccm.
" matico	f ℨ ss to ij	2 to 8 ccm.
" moschi	♏︎xv to lx	1 to 4 ccm.
" myrrhæ	f ℨ ss to j	2 to 4 ccm.
" nuc. vomicæ	♏︎v to xlv	0.30 to 3 ccm.
" opii	♏︎v to xv	0.30 to 1 ccm.
" opii camph.	♏︎v to lxxv	0.30 to 5 ccm.
" phytolaccæ	♏︎v to lx	0.30 to 4 ccm.
" physostigmatis . . .	♏︎v to xv	0.30 to 1 ccm.
" quassiæ	f ℨ ss to ij	2 to 8 ccm.
" rhei	f ℨ j to viij	4 to 30 ccm.
" rhei arom.	♏︎xxx to lxxv	2 to 5 ccm.
" rhei dulc.	f ℨ j to iv	4 to 16 ccm.
" sanguinariæ	♏︎xv to lx	1 to 4 ccm.
" scillæ	♏︎v to lx	0.30 to 4 ccm.
" serpentariæ	f ℨ ss to ij	2 to 8 ccm.
" stramon. fol. . . .	♏︎v to xv	0.30 to 1 ccm.
" stramon. sem. . . .	♏︎v to xv	0.30 to 1 ccm.
" strophanthi	♏︎ij to vj	0.10 to 0.40 ccm.

Remedies.	Dose expressed in terms of apothecaries' weights and measures.	Dose expressed in metric terms.
Tinct. sumbul.	℔v to xxx	0.30 to 2 ccm.
" valer.	f ℨ ss to ij	2 to 8 ccm.
" valer. ammon. . . .	f ℨ ss to ij	2 to 8 ccm.
" veratr. vir.	℔iij to x	0.20 to 0.60 ccm.
" zingiberis	℔xv to lx	1 to 4 ccm.
Trinitrine (nitroglycerine) . .	℔j to iij	0.05 to 0.20 ccm.
Urethan	gr. xv to lx	1 to 4 gm.
Uva ursi	ℨ ss to j	2 to 4 gm.
Verat. alb.	gr. j to iij	0.05 to 0 20 gm.
Veratria	gr. $\frac{1}{12}$ to $\frac{1}{3}$	0.005 to 0.02 gm.
Zinci acet.	gr. j to ij	0.05 to 0.10 gm.
" bromid.	gr. $\frac{1}{2}$ to ij	0.03 to 0.10 gm.
" cyanid.	gr. $\frac{1}{12}$ to $\frac{1}{4}$	0.055 to 0.015 gm.
" iodid.	gr. $\frac{1}{2}$ to iij	0.03 to 0.20 gm.
" oxid.	gr. 1 to x	0.05 to 0.60 gm.
" phosphid.	gr. $\frac{1}{10}$ to $\frac{1}{6}$	0.005 to 0.01 gm.
" sulphas (emetic) . . .	gr. xv to xxx	1 to 2 gm.
" valerianas	gr. j to v	0.05 to 0.30 gm.]

INDEX.

www.ingramcontent.com/pod-product-compliance
Lightning Source LLC
Chambersburg PA
CBHW021802190326
41518CB00007B/413